GOD AND SCIENCE

IN

UNIFICATION

THOUGHT

by

Richard L. Lewis PhD

HYOJEONG ACADEMIC FOUNDATION

For all my Parents

True, Spiritual and Physical

The Publisher

Hyojeong Academic Foundation
34 Down Building 7th Floor
Mapo-gu, Mapo-daero
Seoul City, South Korea

Tel 02-3278-5100

Fax 02-3278-5199

Email admin@wonmo.org

The Author

Questions and comments to:

RICHARDLLL@MAC.COM

Second Edition
ISBN 978-1-387-60000-7

Richard L. Lewis

This book can be purchased at:

https://www.lulu.com/shop

Printed in the United States of America

Contents

INTRODUCTION

This book aims to discourage the view that science and religion are dealing with disjoint realities. To show, to the contrary, that they are often dealing with the same aspects of reality but using very different terms and descriptions.

The introduction to an early version of the *Divine Principle*, compiled by Won Hyo Eu, contained this prescient statement:

> Religion and science have been the methods of searching for the two aspects of truth, in order to overcome the two aspects of ignorance and restore the two aspects of knowledge. The day must come when religion and science advance in one united way.... Then, mutual understanding will occur between the two aspects of truth, the internal and the external.[1]

Such a merging was impossible to imagine even a century ago when religion had a six-day-old universe that science considered an eternal steady state. Things have radically changed, however, with the emergence of sophisticated religious thought—exemplified by *Unification Thought*—and the scientific revolution—exemplified by quantum mechanics and cosmology. These two sophisticated streams of thought are not hostile.

The aim of this book is to explore areas of thought where this convergence of disciplines is apparent, and a final illustration of how the two working together can have constructive dialog. Topics include:

1.   The balance of internal law and external freedom;

2.   The apparent disjoint between a very metaphysical Creator and a very substantial reality;

3.   Internal non-substantial aspects in quantum physics, and the nature of mind;

4.   The anthropic nature of the universe being just right for life;

5.   Two complementary realms in both science and religion;

6.   The underlying cause of life's evolution on Earth;

7.   The duality of digital information and analog form;

8.   The sequence of just-right environments for evolutionary advance;

9.   The Vital Force of evolution

10.   The Origin of Man in science and religion;

---

[1] *Divine Principle* 1973 (Black Book) - Won Hyo Eu

⚜ God and Science in Unification Thought ⚜

11.   Possibilities for Interstellar travel;

12.   Preparations for the burnout of the physical universe

These chapters will examine how science and religion both describe, in their own way, three interacting realms comprising the Cosmos:

1.   An abstract, non-corporeal realm;

2.   A substantial corporeal physical realm;

3.   A substantial corporeal spiritual realm.

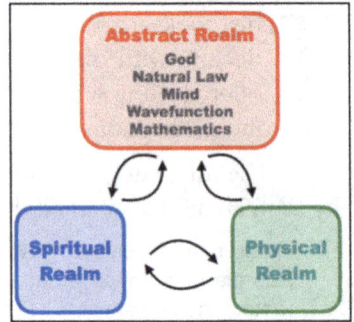

Abstract Realm
God
Natural Law
Mind
Wavefunction
Mathematics

Spiritual
Realm

Physical
Realm

1 • LAW AND LOGOS

T his chapter deals with natural law in modern science and in *Unification Thought* (UT). Both science and UT agree that there was something existing before the Big-Bang birth of the Universe.

For science, it is the mathematics that all theorists use in attempting to explain what caused the Big-Bang. Some would say that Natural Law was already in effect, while believers in the Multiverse—discussed later—assert that these randomly emerged differently in each of the multitude (though there must have been a Law of Laws about their necessity).

Mathematics exists in the Abstract Realm, which pre-existed the universe. The illustration is an illustration of the sophisticated math involved in explaining the Big-Bang[1] which we, thankfully, will not need to explore here.

$$\dot{\rho}+3\frac{\dot{a}}{a}\left(\rho+\frac{P}{c^2}\right)=0$$

$$\dot{\rho}+3\frac{\dot{a}}{a}\rho=0 \;\Rightarrow\; \frac{1}{a^3}\frac{\partial}{\partial t}(\rho a^3)=0 \;\Rightarrow\; \rho \propto a^{-3}$$

$$\dot{\rho}+4\frac{\dot{a}}{a}\rho=0 \;\Rightarrow\; \frac{1}{a^4}\frac{\partial}{\partial t}(\rho a^4)=0 \;\Rightarrow\; \rho \propto a^{-4}$$

For UT, the Creator God designed a hierarchy of natural laws to start the universe and end up with human beings. This hierarchy is called *The Logos* in UT, the *Principle* in the *Divine Principle* and the *Word of God* in the Bible. God and the Logos also exist along with mathematics in the incorporeal abstract realm.

In *Unification Thought*, all of God's creative work went into this abstract construct, the Logos, a concept that embraces, but enlarges, the concept of Natural Law.

Many, if not most, religions embrace the concept of natural law but also insist that God can intervene and do whatever He wants to do without limitation. In UT, God only performs seeming miracles through human beings. Examples would be Moses and Jesus.

Before humans emerged, everything was fully controlled by the Logos alone. In the *Divine Principle*, the period before Man is called the indirect dominion of God. God working through spiritually mature humans is called

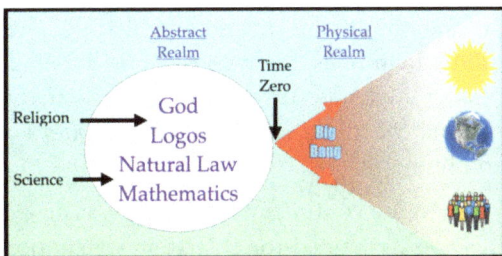

---

[1] https://www.youtube.com/watch?v=b2hdS334QK8

the direct dominion of God.[1] God is not free to ignore the law that He created. The final step in Creation involves Human Responsibility (which has yet to be fulfilled), and God will not take this away.

If science had a dogma it would be: The Universe is fundamentally ruled by Natural Law. While most people assume that scientists view natural law as working at all levels in the hierarchy of science, this is currently not true in the life sciences where Darwinian selection and contingency—not natural law—are thought to govern evolutionary advance.[2]

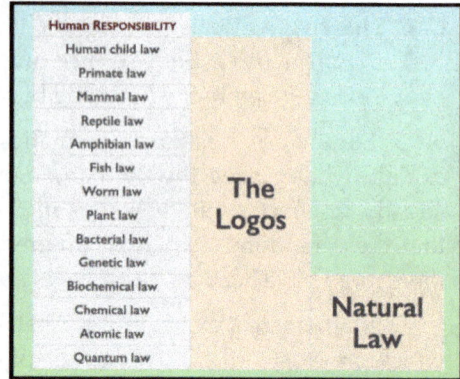

Human RESPONSIBILITY	
Human child law	
Primate law	
Mammal law	
Reptile law	
Amphibian law	
Fish law	
Worm law	**The**
Plant law	**Logos**
Bacterial law	
Genetic law	
Biochemical law	
Chemical law	**Natural**
Atomic law	**Law**
Quantum law	

A simple example involves the right-left isomers of biochemical molecules. All life on earth uses right-nucleotide bases and left-amino-acids—the opposite forms being poisonous to life. Contemporary biology has this as a contingent R-L accident at the origin of life. In this view it could just as well have been R-R, L-L or L-R.

A more sophisticate example is the evolution of multicellular animals. At the early stage of a sphere of cells with hole for food to enter and wastes expelled. A second hole developed for a separation of these two essential processes. These were simple worms—basically a tube of tissue with the original mouth and the emergent anus. These *protostomes* evolved into many different animals, the most successful being the insects. In one lineage, a surprising switch occurred: the first hole became the anus and the second hole became the mouth. These *deuterostomes* evolved into fish, dinosaurs, lions and humans.[3]

In the contingent view, this was an extremely fortunate accident; in UT this was a planned-for change in orientation. The fact of emergent properties is philosophically challenging to materialists, who can only think of it as a lack of ability on the part of humans:

> Because of our inability to directly calculate how complex phenomenon at one level arise from the physical mechanisms working at a deeper level,

---

[1] http://www.tparents.org/Library/Unification/Books/dp96/dp96.pdf#search="divine%20principle%20eu" p. 52

[2] https://science.sciencemag.org/content/362/6415/eaam5979

[3] See "The Worm that Turned" in Christian de Duve's book, *Vital Dust: Life as a Cosmic Imperative*

scientist sometimes throw up their hands and refer to these phenomena as 'emergent'. They just pop out of nowhere.[1]

This contingent and inability aspect is absent from Unification Thought which has the patterns inherent in natural law applying at all levels of the evolutionary hierarchy. The composite of all Natural Laws at every level of nature is called the Logos. This view predicts that all life that emerges in the Universe will be like ours, and thus will not be inherently poisonous.

In a famous thought experiment, eminent evolutionist Stephen Jay Gould asked whether, if one could somehow rewind the history of life back to its initial starting point, the same results would obtain when the "tape" was run forward again. He speculated that it would be very different. To the contrary, UT would have such life basically identical. Only when exobiology becomes a practical discipline will we have the ability to distinguish between Law or Contingency in the origin and history of life.

Hierarchical Systems

In *Unification Thought* the Logos is progressively expressed through evolutionary history. Step-by-step, a hierarchy of increasingly sophisticated systems is expressed in history: atoms progress to molecules to life to plants and animals. The final and complete expression of the Logos in *Unification Thought* is the advent of humans. Unlike the rest of creation, however, humans are not perfected by the power of the Logos alone but have a Portion of Responsibility in their own spiritual maturation and perfection.

At all levels, the basic principle of triple-level system construction applies: 1) All systems are composed of interacting subsystems; 2) these subsystems couple with their sub-subsystems; 3) A system interacts with other systems by coupling with a subset of its subsystems.

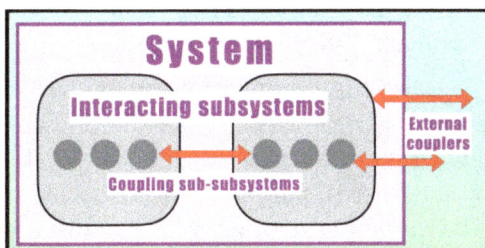

An example would be a nucleotide molecule that is composed of interacting atoms. The atoms interact by coupling with their electron subsystems. The molecule itself can interact by coupling externally with electrons and also with atoms, as in the all-important hydrogen bonding so crucial in biochemistry and genetics.

---

[1] Per Bak, quoted in the *Oxford Book of Modern Science Writing*, editor Richard Dawkins, 2008, Oxford University Press, p. 275.

This hierarchy of systems is developed under the influence of natural law, a resident of the Abstract Realm. Just how an abstract law could govern material objects was a great puzzle in materialistic science. How could abstract principles influence substantial matter? This was elaborated in Nobel Laureate Eugene Wigner's famous lecture *"The Unreasonable Effectiveness of Mathematics in the Natural Sciences."*[1]

Internal Aspect

This influence became less unreasonable when science discovered that there was more to matter than just the external aspect. Early in the 20th Century, an intangible aspect was found necessary in a complete description if the behavior of matter was to be understood.

This non-corporeal, internal aspect is called the *wavefunction* in quantum mechanics, and it can only be delineated by *complex number* math in which linear size and angular rotation are unified into a single measure. This is unlike the familiar external particle aspect that is well-described by *real number* math where size and rotation are dealt with quite separately. The wavefunction is fully discussed in the chapter, *The Wavefunction and the Mind*.

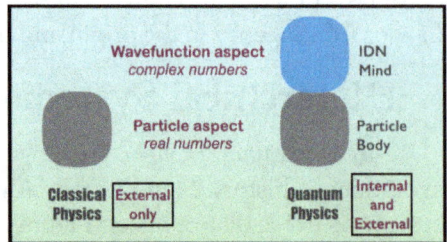

In UT, this aspect of matter is called the *Inherent Directive Nature* of inanimate systems, and the *Mind* of living systems.

In Classical physics, natural law was thought to work directly on the external particle aspect. Quantum physics, to the contrary, found that natural law works directly on the internal wavefunction, and has no direct influence on the external particle. It is reasonable for an abstract law to act on an abstract wavefunction, while to a materialist, it is now the concept of an abstract wavefunction that is unreasonable.

The internal wavefunction is expressed as the probability of how the particle will move and interact. It is a great mistake to confuse quantum probability with classical probability. Quantum probability is

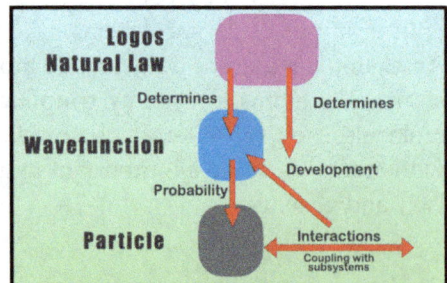

---

[1] *Communications in Pure and Applied Mathematics*, vol. 13, No. I (February 1960). New York: John Wiley & Sons, Inc.

stronger than electromagnetism, stronger than is gravity. So strong that quantum physics embraces the totalitarian principle: **What is not forbidden is compulsory.**

Am example of how different quantum probability is from classical probability (CP) can be illustrated with a pair of dice. Quantum probability comes in two different versions—BP and FP—neither of which is anything like CP. Rather than exploring the technicalities of Bosonic and Fermionic character, a simple analogy will suffice to show their differences to each other and classical probability:

A pair of thrown CP dice can amount to any number from 2 to 12 with 7 the most probable. A pair of BP dice always comes up as doubles—1&1, 2&2…6&6— but never throws a seven. A pair of FP dice, to the contrary, always comes up a 7 —1&6, 2&5…6&1—but never a double. Photons obey BP rules, as in a laser, while electrons obey FP rules, as in the structure of the Periodic Table of Elements.

The external coupling with subsystems alters and develops the internal wavefunction, a give-and-take between internal and external aspects. This development and change in the internal wavefunction by external interaction is governed by natural law. This can be expressed simply as: The internal determines the probability of how the external will move and interact; the external interactions determine how the internal will change and develop.

A well-known example of this is the slit experiment where the wave aspect embraces both slits but, in the interaction with a detector atom, becomes localized around it. While this probabilistic aspect to natural law might seem esoteric and solely of interest to physicists, there is a simple everyday occurrence that completely flummoxed the genius of Issac Newton and was only explained completely by the advent of the internal wavefunction in quantum mechanics.

This familiar phenomenon is your partial reflection in a shop window while seeing the mannequins on display inside. Each photon of light passing through the glass has a small probability of bouncing back, so a small number of the billions passing through the glass are returned to you. If you want to know more about this internal aspect, I recommend Richard Feynman's *QED: The Strange Theory of Light and Matter* as a great introduction for the nonspecialist.

Nobody can predict what a photon, or any other particle, will "choose" to do, or how it makes the choice, a problem that vexes philosophers to this day.

Determinism and Free Will

UT, along with the Abrahamic religions, has always asserted that humans have free will, and are responsible for their actions. Materialistic science, however, had a problem with this as it embraced determinism: that natural law governed all things, including the human brain, so behavior was determined by chemistry, not free will.

This implication of materialism is problematic in both the justice system and organized religion where people are held responsible for their actions. This does not seem reasonable if criminal actions are just the result of chemical imbalances in the brain. If so, perhaps chemical adjustment should be imposed to redirect the criminal chemistry at work.

Fortunately, modern science is no longer simply deterministic. The view of the relation between abstract natural law and what actually happens externally is now more complex and sophisticated. While the development of the internal wavefunction by interaction is fully determined by natural law, the internal law only determines the probability of what the external will do: what actually happens is not determined, it is probabilistic.

That elementary entities have an element of freedom seems unreasonable to materialists. It should be noted that this random choice—in the sense that it is not determined—amongst a set of probabilities is the most difficult thing to emulate in computers, which almost by definition, are defined by their programing. The best they can do is generate a sequence of numbers whose properties approximate the properties of sequences of random numbers.

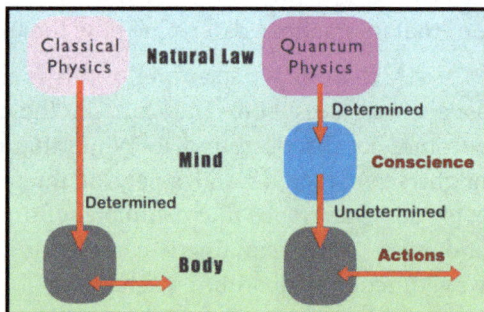

Unlike post-Reformation science, Quantum science has plenty of room for free will. In human experience, the natural law of the Logos determines that our conscience directs us to do good, but we are free to ignore this and do something else.

Emergent Properties

All entities have an internal wavefunction; their external interactions by coupling alters this wavefunction, as determined by the Logos. The altered wavefunctions of the entities when interacting as subsystems of a system merge together to generate the wavefunction of the entire systems. This is the origin of the wavefunction of every system, from the simplest—such as the atom—to the most sophisticated—the human brain.

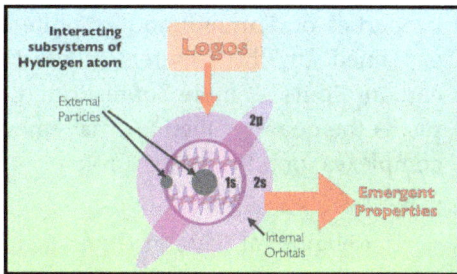

The emergent wavefunction has a set of properties determined by the Logos. When simple systems interact and combine as subsystems into a more sophisticated system, the emergent system displays a set of characteristics that are entirely absent in the subsystems. These "emergent properties" appear at every level of sophistication. The emergent properties come from the Logos; states UT; they just happen from nowhere according to modern science.

An example: neither a proton nor an electron have the property called 'chemical valence.' They both have an external particle aspect and an internal abstract aspect. Note: Science is so unsure about the internal aspect of matter that it gives it different names at every level, which can be confusing: *Probability amplitude* for a particle moving from one place to another; *Wavefunction* for the overall behavior of simple entities; *Atomic Orbital* for electrons in atoms; *Molecular Orbital* for atoms and electrons in molecules. It can be confusing.

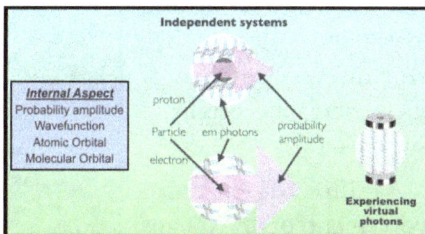

While the proton has a complicated internal structure and the electron is called a "fundamental" particle, both particles are composite, their subsystem structure including a cloud of virtual photons called an electromagnetic field. (If you find it hard to believe in non-real photons, try pressing the North poles of two strong magnets together: the invisible cushion you feel between them is composed solely of virtual photons.)

The proton and electron particles couple with these virtual subsystems. This alters the internal aspect of both, governed by the Logos, which merges them into an atomic wavefunction called an orbital. This resultant orbital gives the nascent system, called a hydrogen atom, the emergent property of chemical valence.

This principle applies throughout the hierarchy of matter: the external subsystems retain their individual identity while their wavefunctions merge into the unified entity of the system wavefunction.

Just where these emergent properties come from is not a question asked, or answered, in modern science. All it can say is that the contrasting properties of diamond and graphite can be explained by the different interaction of carbon atoms. A more sophisticated example, is the quality called *life* that emerges in complexes of interacting proteins, nucleic acids, etc.

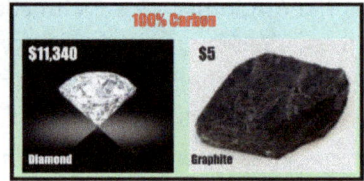

Religion, at least in *Unification Thought*, states that these qualities are imbued into natural law by the Creator and—according to science—are expressed via the wavefunction into the physical realm. This is similar to the quality of maternal sorrow and grief expressed in marble as the *Pieta*; these emergent properties, not possessed by a block of marble, come from the mind of Michelangelo.

In UT, it is the input from the Logos that drives the process of evolution (see later chapter) and is the source of the emergent properties, while in Darwinism, these new characteristics, such as life and thought, appear out of nowhere, randomly by chance and accident.

Living and Inanimate Systems

While there are a great number of differences between living and inanimate systems, we will discuss just one of them here; the origin of systems.

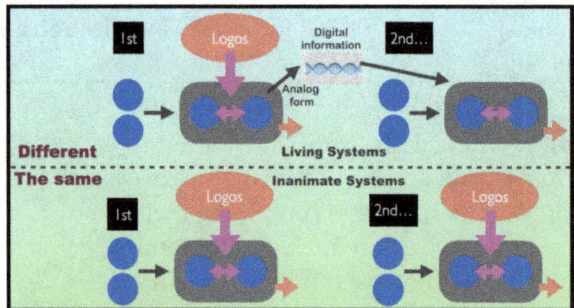

The direct involvement of the Logos applies to the emergence of all non-living systems. For example, the origin of every hydrogen atom is the same as the origin of the very first—and there was a first—hydrogen atom in the universe.

This does not apply to living systems. The origin of the very first of a species is exactly the same as in non-living systems—a new analog pattern of interacting subsystems from the Logos is expressed along with its set of emergent properties. In living systems, however, this novel analog pattern from the Logos is captured in digital information that is stored in the nucleic acids.

The analog form of the second, third, etc, generations does not involve a direct input from the Logos but is directed by the stored digital information passed down the generations in the DNA. The emergent properties of the analog forms are still, however, a reflection of those in the Logos.

While the origin of the first of any system involves the direct input from the Logos, the origin of the second, and all subsequent living systems is very different as these have input from stored digital information rather than direct input from the Logos.

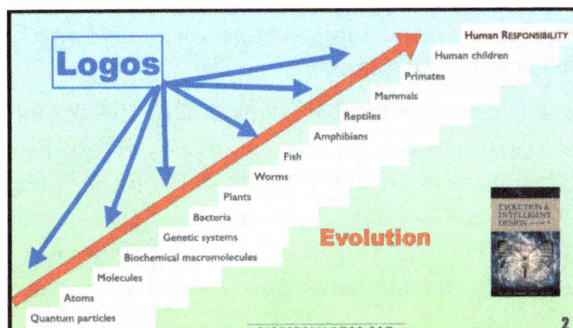

Properties from the Logos

The table below is a selective list of some of the emergent properties of systems, with the Logos expressed in the analog form of the interacting subsystems and functioning of systems.

Interacting subsystems	Resultant System	Emergent Properties from the Logos
Protons & electrons	Atom	Chemical valence
Atoms	Molecule	Hydrogen bonding
Amino-acids	Proteins	Manipulation of analog form
Nucleotides	DNA, RNA	Manipulation of digital information
DNA, RNA, Proteins	Cell	Life
Glia & Neuron cells	Brain	Human mind

Hydrogen bonding is of paramount importance in the structure and function of proteins—with their extraordinary ability to manipulate the analog form of molecules—and the nucleic acids—with their exceptional ability to manipulate digital information.

The interactions of these macromolecules express the quality of life from the Logos in simple cells, such as bacteria, and the variety of cells in the plant and animal world.

The interactions of cells underly the Logos-derived functioning of all the organs in the human body. Of particular note being the quality of the human physical mind expressed through the interactions of the glia and neuron cells—with complementary roles similar to DNA and proteins in cells—that compose the human brain.

In this way, the Logos drives the evolutionary process. The many books on Intelligent Design make an excellent case for the failures of Darwinism random mutation, and evidence of an input by intelligence into evolution. They do not, however suggest just how this input occurred in history. Unification Thought, in the manner just described, does suggest an outline for the mechanism by which this input occurred.

This is detailed in a later section.

The action of natural law—the Logos—is not remarkably in the quotidian details of everyday life. We do take it for granted. It is, however, noticeable when a new emergent character appears on the scene of history. We shall just mention a few here.

PLASMA MEMBRANE

All known living systems are is separated from the environment by a bi-lipid membrane. The aqueous outside and inside are separated by two layers of linear molecules with water loving and a water hating ends. They have their water loving ends on the inside and outside, and the water hating ends layered in between. They also had a single origin:

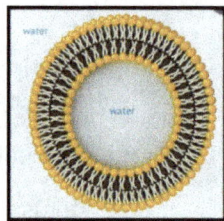

> Membranes grow by accretion, that is, by the addition of components to a pre-existing membrane. Thus, de novo synthesis of a membrane needed to occur only once in the history of life, and all subsequent membranes could have just arisen from this ancestral membrane by expansion followed by fission.... At least, membranes develop in this manner in the living world today.[1]

This (unknown) origin event was instigated by the Logos.

LAST UNIVERSAL COMMON ANCESTOR

The current consensus is that all of "Our Type of Life" (OTL) is descended from a single ancestor who bequeathed to us all a set of properties we hold in common. These include the universal genetic code (digital to analog conversion), the Krebs cycle (manipulation of carbohydrates), and ATP (the universal carrier of chemical energy). Less well known is chemiosmosis,

---

[1] Christian de Duve, 1995, *Vital Dust: Life as a Cosmic Imperative*, BasicBooks, NY, p. 93

universal to all life, the use of acid gradients to drive metabolism. All these appear to be in properties of the last universal common ancestor, LUCA.

This was the ancestor of the two extant types of bacteria—the eubacteria and the archaebacteria,. As explained by a non-believer in the Logos:

> In a fluke of fortune bordering on the unbelievable, it might be that both the [two types of bacteria] emerged from the very same hydrothermal mound. Little else could explain the fact that they share the same genetic code, as well as many details of protein synthesis, but apparently only learnt to replicate the DNA later on, totally independently.... The evidence that all the living organisms descended from a single common ancestor is overwhelming. We cannot exclude the possibility that the unknown or poorly known organisms of a different origin exist in some remote environment that has remained isolated for a very long time. However, no discovery suggestive of a major break with "our" way of life has yet been made. Until proven otherwise, the hypothesis of a single ancestry holds true.[1]

Understanding the role of the Logos removes this crucial period from the 'border of the unbelievable.' As we will see later, this activity of the Logos generating two types of bacteria was crucial to the emergence of non-bacterial life, including the human.

ORIGIN SOPHISTICATED CELL

The transition from simple bacteria (prokaryotes) to sophisticated cells (eukaryotes) was a new emergent property from the Logos:

> Complex life did not emerge repeatedly from bacteria at separate times. Plants from one type of bacteria, animals from another, fungi and algae from yet others. On the contrary, on just one occasion a complex [entity] arose from bacteria, and the progeny of this cell went on to found all the great kingdoms of complex life: the plants animals fungi and algae.... The eukaryotic cell only evolved once because the union of two prokaryotes, in which one gains entry to another, is truly a rare event a genuinely faithful encounter. All that we hold dear in this life, all the marvels of the world, stem from a single event that embodied both chance and necessity.[2]

"Chance and necessity" being the scientific vague understanding of the Logos at work.

THE WORM TURNS

One of the most odd events in biological history is the origin of the deuterostomes. It is extraordinarily difficult to explain this transition as a chance and accident occurrence. This description is from a Nobel laureate

---

[1] Lane, Nick. 2009, *Life Ascending: Ten great inventions of evolution*. W. W. Norton New York, p.31-35

[2] Lane, Nick. 2009, *Life Ascending: Ten great inventions of evolution*. W. W. Norton New York, p.31, 55

darwinist who makes no attempt to explain this "accident" in the lineage of simple worms:

> The cells arising from the early divisions of the fertilized egg first form a sphere... which turns into a double walled pouch... with a single opening the blastopore.... [it] acquires a second opening and turns into a canal. [In all simple animal lineages] the blastopore becomes the mouth and the new opening the anus. They are called the protostomes (mouth first) for this reason. [Earthworms, insects, mollusks, etc.]

> The historic flip that started a new line made the blastopore the anus and the new opening the mouth thereby initiating the deuterostomes [mouth second].... It is possible that, without this fateful switch, there would be no fish, no amphibians, no reptiles, no birds, no mammals, no humans.[1]

It was the action of the Logos that made this historic 'worm turning' a high probability, rather that a pure accident. We will continue up the levels of sophistication as the discussion develops.

---

[1] Christian de Duve, 1995, *Vital Dust: Life as a cosmic imperative,* BasicBooks, NY, p. 199

2 • METAPHYSICAL CREATOR, PHYSICAL CREATION

G od and mathematics have at least one thing in common: They are everywhere; as big as the Universe, as small as a proton. Try to convince scientists that the proton is too small, or the Universe too big, for mathematics to apply there: they will have you committed. The same would happen with theologians. Both God and mathematics are beyond the physical; they are metaphysical.

A very natural question is: How can a metaphysical being create a physical Universe? If Julie Andrews, in the *Sound of Music* is correct, that: ♫ *Nothing comes from Nothing; Nothing ever could,* ♫ then we have a conundrum because the universe is full of physical things and God is not one of them.

The answer we will be scientifically exploring in this book is: There is no fundamental difference between metaphysical and physical entities.

To start, we note: All sandcastles are fundamentally the same; shaped sand. They can be as very different as a child's to an artist's. These constructs have one type of matter—the sand grain—and one type of force to hold them together—water.

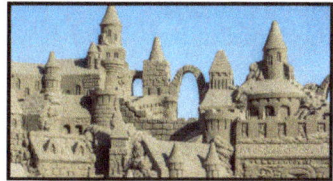

One of the truly astonishing discoveries of science in the last century is that the physical world is almost as simple as one particle and one force. It is firmly established that in the physical world there are just four fundamental matter particles and four fundamental forces binding them together.

The four forces are: **gravity**, which holds us to the ground; **electromagnetism** which runs everyday life; the **strong** nuclear force, which powers the sun; and the **weak** nucleus force which stops the sun from exploding all at once, keeping it burning sedately for billions years.

The four fundamental types of matter are: the **neutrino**, which we do not notice even though billions of them zip down through us from the sun at midday and the same number zip up through the earth at midnight; the featherweight **electron,** so important in the biochemistry of our bodies; and the **U quark** and the **D quark** of the atomic nucle-

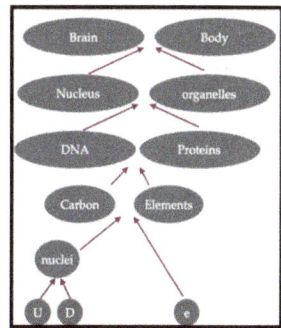

us, the source of almost all the mass of matter.

As far as humans are concerned, our external self is hierarchically constructed from these simple components.

Spacetime and Gravity

One of the great mysteries remaining in science is: What is spacetime, the empty vacuum. Science knows somethings about how it was created. Ignoring the details: A tiny speck of False Vacuum was created in the metaphysical realm. Its size was a Plank Length (10^{-33} meters) and at the Planck Temperature (10^{36}K) and density (10^{93} gm/cc) and packed with every kind of particle, including the quarks. This doubled in size every Planck Time (10^{-44} second).

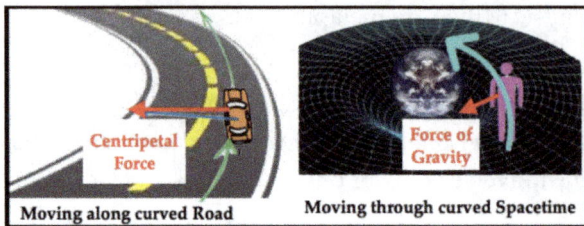

Moving along curved Road   Moving through curved Spacetime

It took 10^{-28} seconds, however, for the quarks to respond to their separation with an immense burst of energy (to be shortly discussed) generating the Hot Big-Bang, and slowing the inflation rate considerably. By this time the seed universe had inflated to $2^{10^{16}}$ a vast inflation of its it original size. This was the creation of spacetime, and it has been expanding—much, much slowly—ever since.

No one has been able to say what spacetime actually is, but it has a set of abstract properties:

• It has four components, all at 90° to each other

• One of time is measured in real numbers, three of space are measured in imaginary numbers[1]

• It has the property of separation given by the Pythagorean relation (the metric). More details later.

• Spacetime can bend—the phenomenon underlying universal gravity. (This is an excellent introduction to this weighty topic[2])

• Spacetime is expanding

---

[1] It used to be the other way around

[2] Zee, A. *On Gravity: a brief tour of a weighty subject*, Princeton University Press. 2018

• Everything moves at lightspeed through spacetime, although the components can differ.[1]

Note: This implies that a second is a vast amount of time, equivalent to 186,000 miles of space! It is this universal velocity through curved spacetime that, like that experienced in a speeding car traveling around a curve, is the universal force of gravity. A photon, of light traveling at lightspeed along the spatial dimension, and a human, traveling at lightspeed along the temporal dimension, both experience the same force of gravity.

Einstein came close to this viewpoint:

Imagine a coordinate system which is rotating uniformly with respect to add initial system in the Newtonian manner. Centrifugal forces which manifest themselves in relation to the system must, according to Newton's teaching, be regarded as affects of inertia. But these centrifugation forces are, exactly like the forces of gravity, proportional to the masses of the bodies. Ought it not to be possible in this case to regard the coordinate system or stationary and the centrifugal force is as gravitational forces?[2]

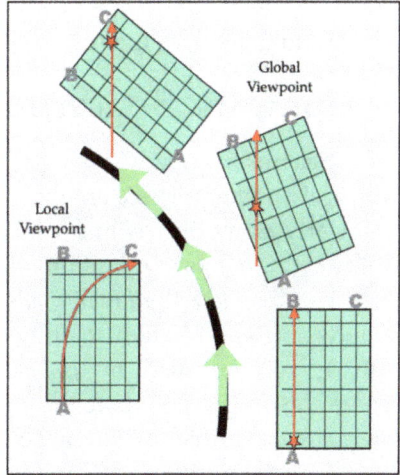

The curved-road analogy without the car's wall keeping one in place is also illustrative. In a thought experiment: An ice rink is being trucked along a curved highway. A skater starts off at point A and heads straight towards point B without changing direction. We see from the global point of view that as the truck rounds the curve, the ice rink rotates, but the direction of the skater on the frictionless ice does not. The skater who aimed at B ends up at C and concludes that in his local reference frame there is a forces of attraction pulling on him.

When asked, however, if he had felt a force pulling him, he said he did not. This is akin to what Einstein said, "For an observer falling freely from the roof of a house, the gravitational field does not exist". He meant that literally, and later described it as "the happiest thought of my life".

Experimentally-minded, the skater, a man of 300 lbs, asked his 50 lb daughter to skate with him, and they both ended up at point C. He concluded that the force is independent of sex and weight. Finally, he asks the truck

---

[1] Greene, B. *The Elegant Universe*, 1999, p.50

[2] Quoted in, *The Oxford Dictionary of Modern Science Writing*, editor Richard Dawkins, 2008, Oxford University Press, p. 317

to speed up so that point C moves down the side until a speed is reached such that even though he aims across the rink he ends up moving along the side he started on, he cannot leave that side no matter how fast he skates. He is trapped as in a Black Hole!. In this analogy, the speed of the truck is the mass, while the constant speed of the skater is lightspeed. When the truck stops (no mass) the skater aims at and reaches point B

So spacetime is an insubstantial entity that can be curved. For this discussion to proceed, we will need two more properties to spacetime, properties that explain a great deal but are currently just speculations:

- Spacetime can be twisted

- Rectangular spacetime can be deformed into hexagonal spacetime.

Complex Spacetime

Scientists initially attempted to describe the world mathematically with *real* numbers. These are the usual numbers that have a size that stretches linearly along the *real* line from minus infinity through zero to plus infinity. Modern science has realized that reality also needs to include angular rotation to fully describe it. This is accomplished by *complex* numbers where the real line is rotated by 90° into the *imaginary* axis, denoted by *i*, generating the complex plane. Complex numbers on this plane have both a linear size and an angular rotation.

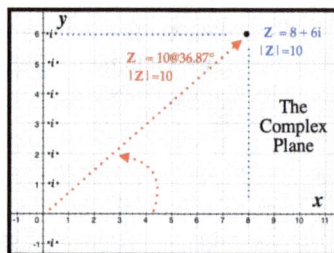

Modern quantum physics has established that all entities, including spacetime, have dual aspects: There is an internal aspect described by these complex numbers; This has an external extension, described by real numbers, that we call probability. The internal complex value is always expressed externally as its absolute square that is always positive. While plus and minus *real* numbers are very different, plus and minus *imaginary* numbers are relative, like clockwise and anticlockwise which depend on the viewpoint.

The four components of spacetime divide into three spatial components—traditionally notated as *x, y, z*—and one temporal component—traditionally notated as *t*. In the early days, time was thought to be described by *imaginary* numbers, while space was described by *real* numbers. This has now changed with the realization that the past and future are absolutely different, as are plus and minus real numbers; while up and down, forward and backward, here and there, are all relative, as are imaginary numbers. This is why time is considered mathematically *real* while space is mathematically *imagi-*

nary. While it is possible for particles to move in either direction in time on an internal level—matter and antimatter—they both move in the same direction externally as the square of a *real* number is always positive.

This internal aspect of spacetime also allows for the rather unintuitive concept that an entity can exist on the internal level without any external aspect at all. These are called virtual particles, and they do not have energy-in-time that amounts to a real pixel of existence.

Twisted Quanta

We will now explore the consequences of the spatial components of spacetime being twisted around the temporal axis. There are two types of twists in the spatial components of spacetime; symmetrical and asymmetrical. Each has an associated complex waveform and boundary. The energy of the wave being the square of the amplitude—always a positive number.

A symmetrical twist of a spatial component by 360° around the time axis does not alter the component, and a transition of an entity through the twist leaves its state unchanged. The technical term is a "boson spin of 1". The waveform is the cosine wave with all its energy—cos²—at the center and boundary. This is an open wave with a high-energy boundary.

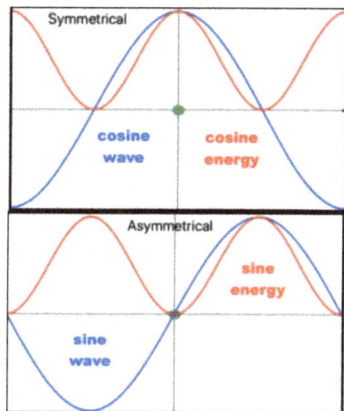

An asymmetrical twist of a spatial component by 180° around the time axis mixes the component, and a transition of an entity through the twist flips its state which can be restored by a second transition. The technical term is a "fermion spin of ½". The waveform is the sine wave with zero energy—sin²—at the center and boundary. This is a closed wave with a zero-energy boundary.

The word *quanta* comes from the Greek, and in science means a discrete amount of energy. The energy of both matter and forces come in quanta. The energy of these quanta is in the localized twists in the spatial components of spacetime. The quanta of forces are symmetrical twists, the quanta of matter are asymmetrical twists. The twists can be in one, two or all three components; so there are three types of force quanta and three types of matter quanta.

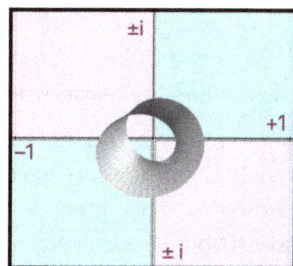

We will first look at the broad outline of these quanta, then mention any complicating details at the end.

ONE TWISTED COMPONENT

A twist in one spatial component—call it the **x** component—is called quantum spin, and it can be asymmetrical ½-**spin** or symmetrical **1-spin**.

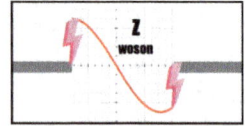

The symmetrical single twist is called the Z weak boson—we will abbreviate to Z woson—with a spin of 1. Its localized high-energy open boundary generates so much stress in spacetime that the Z woson has the enormous energy of 91,000,000,000 eV (91 GeV). This enormous energy is reduced somewhat to 80.4 GeV by generating and resonating with an electron—a W⁻ woson—or a positron—a W⁺.

The single asymmetrical twist in the x component of space is called the neutrino with a left-handed spin of ½. (For reasons unknown, the only difference between the neutrino and antineutrino is the direction of it spin, antineutrinos spinning

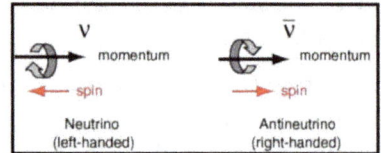

right.) Its zero-energy boundary hardly stresses spacetime at all, and it has so little energy that it has not been precisely measured but is in the range of ~1 eV. Nevertheless, spacetime resists this twist and attempts to shake it off. This is topologically impossible, and all that results is a 1-twist woson.

There is no energy available for one, so the neutrino is surrounded by a halo of virtual wosons that briefly flicker on and off. Even a virtual woson has temporary mass so it does not travel far—even at lightspeed—before disappearing. This weak field is very small, even on the scale of a proton. This halo is called the "weak charge" on the neutrino.

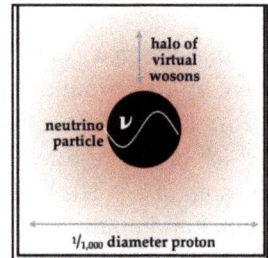

TWO TWISTED COMPONENTS

The twist in one spatial component—call it **x**—is called magnetic spin and the twist in the second component—call it **y**—is the electric vector.

A force quanta with two symmetric twisted components is called a photon with a spin of 1. The open boundary is not a local stress on spacetime as the energy is constantly shifting from one twist to the other—the electromagnetic wave—following the rule that a changing electric field induces a magnetic

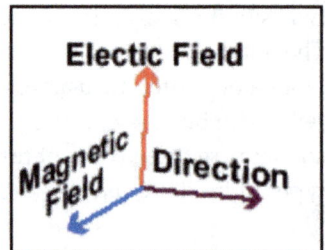

field, while a changing magnetic field induces an electric field.

They waves are related as cosine and sine and the sum of their squares—the energy—is a constant. This waves moves at light speed along the third, z, spatial component. The energy of a photon, E, is all in the wave itself which depends on the time period, T. Energy over time is called **the action** in physics, and each photon has a single quanta of action, where ET=1 in the appropriate units. Radio waves have a big T and small E, while gamma rays have small T and big E.

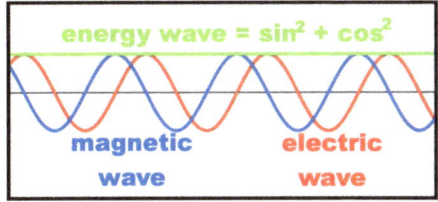

energy wave = $\sin^2 + \cos^2$

magnetic wave   electric wave

A matter quanta with two asymmetric twisted components is called an electron with a spin of ½. Having two localized twisted components does stress spacetime, and the electron has rest energy of 500,000 eV.

Spacetime attempts to shake off these double twist, which is topologically impossible, and the result is a halo of virtual force particles. The electron has a weak charge—a halo of virtual wosons, and an electric charge—the halo of magnetic spin generated virtual photons. A positron is very similar to an electron but it is going backwards in time on the internal level, and its virtual photons are polarized in the opposite directions—it has a positive electric charge—and the opposite magnetic alignment. Meeting each other, the two ½ twists cancel each other and a burst of energetic real photons is generated.

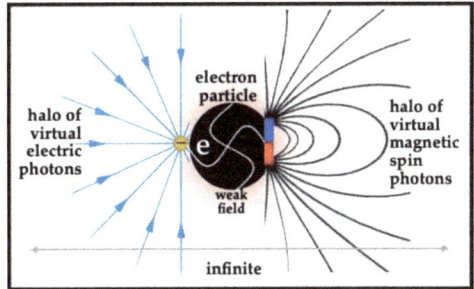

electron particle

halo of virtual electric photons

halo of virtual magnetic spin photons

weak field

infinite

THREE TWISTED COMPONENTS

The twist in one spatial component—call it x—is called magnetic spin; the twist in the second component—call it y—is the electric charge; the twist in the third component—call it z —is the chromodynamic *color* charge. There are three positive color charges, RGB (like those that can color the black computer screen white); and three negative color charges, CMY, (like those that can color white paper black). These quantum colors actually have nothing to do with the colors we see around us.

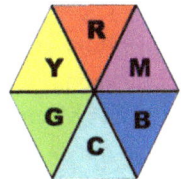

A matter quanta with three asymmetric twists is called a quark. The local stress on spacetime is so great that spacetime is deformed from a 3-D rectangular form of 90° to a 2-D hexagonal form of 120° with time at a right angle to the 2-D plane. One of the components remains unaffected as the spin axis, and there are three possibilities—x, y, z—so there are three *colors* to quarks—**R, G, B** —labeling the unchanged axis.

The second component, on the same plane as the first, is deformed by 30°, while the third is deformed 90° onto the plane and 30° on the plane, 120° in total. There are two planes possible, with the y or the z and the electric compo-

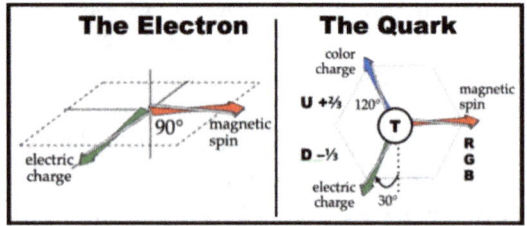

The Electron | **The Quark**

nent can be on either one. So there are two flavors of quarks with different electric charges, the D with –⅓ and the U with +⅔.

Naturally, spacetime vigorously resents this deformation into a planar hexagon form with two consequences:

• The deformation must be strictly confine to a very limited region. This is called the confinement of quarks.

• The color deformation has to be neutral at a larger scale—the deformations have to cancel out—which is why matter quarks have to come in colorless triplets of RGB.

• The deformation of spacetime must be short enough not to amount to a quanta of action, so the quarks are constantly shedding their color, but topological restraints insist that another color be there.

GLUONS
color

Anti-color

This constant shedding is the constant generating of colored gluons, the quanta of force A gluon is an open wave with a color at one end and an anti-color at the other; the center is uncolored. All the energy is at the ends. There are eight gluons, only 3 of them are illustrated here.

The quarks shake off their color so rapidly—changing from one color to another—that all the color ends up in the gluons, and the quarks at the center behave as if they were uncolored, with just electric charge and magnetic spin. The colored and anti-colored ends of the halo of gluons surrounding the quarks speckle the surface of the nucleon as pixels, and from a distance the surface looks uncolored. Similarly, at high magnification, my

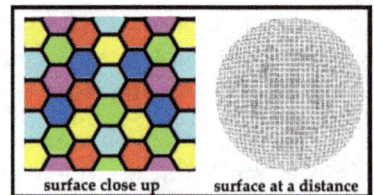

surface close up | surface at a distance

computer screen which seems white is actually composed of tiny pixels of red, green, and blue.

A colored disco ball is the only illustration that comes to mind for the surface of, say, a proton, with the three quarks rattling around in the colorless center.

The colorless situation can also be attained by a quark and antiquark, a meson, and the surface is now pixels of a color and anticolor, colorless black rather than white. It is such virtual mesons that couple the protons and neutrons together in the strong nuclear force.

All the interaction energy of the quarks—which is considerable—ends up in the surface and its gluon ends. The disco ball is hollow, and its skin is thick lead. A quark, perhaps hit by a high-energy electron, that attempts to escape the trio, will generate energy that increases by the 6th power, more that enough to generate a whole jet of newly created particles. An isolated colored quark (or gluon) with its hexagonal nature is impossible.

Fermion	weak	elc.mag	color
neutrino	✔		
electron	✔	✔	
quark	✔	✔	✔

It was this generation of energy when the original quarks were almost instantaneously separated by inflation that caused the Hot Big-Bang and stopped the exponential inflation.

Interaction

Each type of half-twist fermion is surrounded by a halo of whole-twist bosons.

- The single twist fermion (neutrino) is surrounded by a halo of single twist bosons (wosons). It has said to have a "weak charge".

- The double twist fermion (electron) is surrounded by a halo of single twist bosons (wosons) and double twist bosons (photons). It has a "weak charge" and an "electromagnetic charge".

- The triple twist fermion (quark) is surrounded by a halo of single twist bosons (wosons), of double twist bosons (photons), while triple twist bosons (gluons) are strictly confined. It has a "weak charge," an "electromagnetic charge" and a "chromodynamic-color charge".

Fermions interact with each other by coupling—transfer, exchange or sharing—with their halos of complementary bosons.

- Coupling with single wave wosons is the "weak interaction." All fermions can do this.

- Coupling with double wave photons is the "electromagnetic interaction". The neutrinos cannot do this.

- Coupling with gluons is the "color interaction." Only quarks can do this. The "strong interaction" that holds protons and neutron together in the atomic nucleus is an offshoot of this involving coupling with virtual quark-antiquark pairs (pions).

Three Generations

The physical world is fundamentally made of electrons, U quarks, and D quarks interacting by coupling with photons and gluons. The neutrinos and wosons do not have a role in everyday life, except for their extremely significant role in moderating the Sun's burning and preventing it from exploding all at once rather than burning steadily for billions of years.

For the symmetrical quanta of force, there is little else to add, while the asymmetrical quanta of matter are more complicated. These come in three generations.

	1st Gen	2nd Gen	3rd Gen
1 twist	e-neutrino	μ neutrino	τ neutrino
2 twists	electron	muon	tauon
3 twists	U-D quarks	S-C quarks	T-B quarks

The *flavors* of matter quanta we have described so far—electron-neutrino, electron, U-quark, D-quark—all belong to the 1st generation.

There are two other generations; the second generation being muon-neutrino, muon, C-quark and S-quark; the third being tau-neutrino, tauon, B-quark, T-quark. The rather klutzy nomenclature being a residue of the difficult history establishing its details. These 2nd and 3rd generations play little known role in everyday life and are unstable, but they have being uncovered in high energy experiments in the laboratory.

The first generation can be founded on the electron neutrino, ν_e, with one twist in the x spatial component, adding a second twist is the electron, adding the third twist are the U and D quarks. When the first twist is in the x & y components, it creates the muon neutrino, the foundation of the 2nd generation; adding extra twists give the muon and the C, S quarks. When the first twist is in the x, y & z components, it creates the tau neutrino, the foundation of the 3rd generation; adding extra twists give the tauon and the B, T quarks.

Insubstantial Universe

Spacetime is real, but incorporeal. Scientists are still trying to understand spacetime and its properties, exploring high-energy untwisted excitations of the vacuum such as the Higgs boson. It has been said that physics these days is attempting to understand what nothing is.

God is incorporeal, and all things—both matter and forces—are abstract distortions of incorporeal spacetime: Gravity is curved spacetime, matter is asymmetric twists in spacetime, the three quantum forces are symmetrical twists. In this way the apparent distinction between an incorporeal God and a corporeal universe disappears: God didn't make any solid matter, He just twisted spacetime in about a dozen ways to result in all the delights of creation.

Some have even gone so far as to speculate that reality is all mathematical, that all things are mathematical manifestations. That Creation was a mathematical manipulation and "How to Build the Universe using only Math."[1]

Julie Andrews was wrong: A thing <u>can</u> come from no thing, if nothing is sufficiently clever enough.

---

[1] Manil Suri, 2022, *The Big Bang of Numbers: How to Build the Universe using only Math,* Norton NY.

3 • THE WAVEFUNCTION AND THE MIND

T here are basically three different ways to approach a great movie such as Star Wars IV:

1.   Wow, I loved it!

2.   What's the big deal?

3.   How did they do that?

In this book, we are not going to be enthralled by God's Creation, or take it for granted, rather we are going to ask: How did Heavenly Parent construct the world, the realm of scientific curiosity. Hopefully, this talk will be comprehensible to those with a science background. For those not so fortunate, I hope you take away the sense that science and religion are in agreement on this topic.

In the previous section we discussed the Logos, natural law, and how it was expressed through the wavefunction in the physical world. In this book we will discuss the wave function in detail.

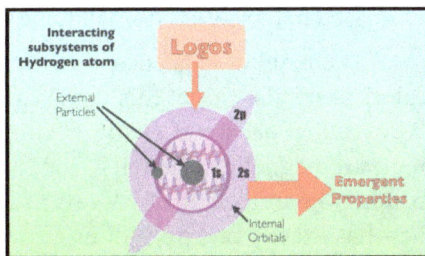

When science really got started about a demi-millennium ago, a great deal of effort went into exploring the various laws that governed the many different aspects of nature: gravity, light, magnetism, fluid flow, etc. Nowadays things are a lot simpler, if more sophisticated. Scientists want to understand how to calculate just one thing about the world; *The Action*.

The Action

A noted scientists explained this development in nontechnical terms:

"Our search for physical understanding boils down to determining one formula. When physicists dream of writing down the entire theory of the physical universe on a cocktail napkin, they mean to write down the action of the universe. [It would take a lot more room to write down all the equations of motion]… The action, in short, embodies the structure of physical reality.... Some physicists would like to believe that the Ultimate Designer thinks in terms of the action."[1]

While the equation is complicated and takes graduate years to understand, anyone can see the many +'s adding up all contributions of all the interactions to get the total action.

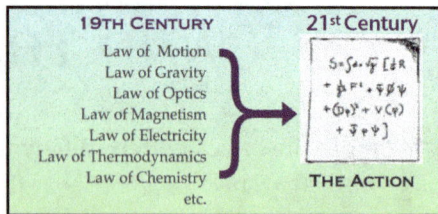

19TH CENTURY	21st Century
Law of Motion Law of Gravity Law of Optics Law of Magnetism Law of Electricity Law of Thermodynamics Law of Chemistry etc.	$S = S_o - \frac{1}{G}[\dot{c}R$ $+\frac{1}{4}F^2 + \bar{\psi}\not{D}\psi$ $+(D_\mu\psi)^2 + V(\varphi)$ $+\bar{\partial}+\psi]$ THE ACTION

The reason that scientists are on this quest is that a very simple rule seems to govern all that happens in the world: The Principle of Least Action. The Action is not a familiar term to nonscientists, for it is the scientific measure of existence, with the units of energy over time.

Intriguingly, beneath the very different ideas about the universe found in Newtonian physics, Maxwell's electromagnetism, quantum mechanics and Einstein's relativity, we find a common core. They are all boil down to a principle of least action.[2]

Thou Shalt Minimize The Action

Like almost everything that was considered continuous in classical science existence—like time, space, energy—actually comes in discrete pixels, called quanta. All of these are grainy, but seem continuous because, like the 326 dpi pixels of my computer screen, they are too small to be noticed.

The pixel of existence, called a Planck's Constant, is tiny at 1.3×10^{-34} Calorie-seconds. It got this name as Max Planck (1858-1947) was the first to discover that each and every pixel-quanta of electromagnetic radiation—from radio, to microwave, to light, to X-ray to gamma—has exactly one pixel of existence, of the action. Light minimized its action by always taking the path of least time. There are also some simple entities in the world that do not have even this tiny pixel of existence, called virtual particles. They are the underpinnings of the electric and magnetic fields.

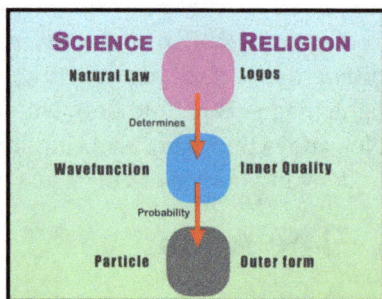

SCIENCE	RELIGION
Natural Law	Logos
Determines	
Wavefunction	Inner Quality
Probability	
Particle	Outer form

[1] A. Zee, *Fearful Symmetry*, Macmillan, NY (1986), pp. 106 - 112.

[2] G.F. Lewis, L.A. Barnes, *A fortunate universe: Life in a finely tuned cosmos*, Cambridge University Press, 2016 p.228

This basic law commanding minimizing the action underlies all that happens, and it corresponds to the Universal Prime Energy in the *Divine Principle*:

> "The fundamental energy of God spirit is also eternal, self existent, and absolute. It is the origin of all energies and forces that allow created beings to exist. We call this fundamental energy universal prime energy. Through the agency of universal prime energy the subject and object elements of every entity form a common base and enter into interaction. This interaction, in turn, generates all the forces the entity needs for existence, multiplication, and action."[1]

Probability Amplitude

While the action is simple in principle, its application can be tedious. Consider an electron and our desire to know if it will go from y to x. First we have to consider all the possible paths it by which it might make the transition.

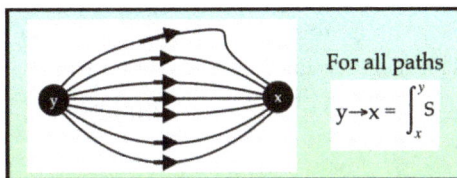

For all paths
$$y \to x = \int_x^y S$$

Then for each path we have to integrate the action over the path. The path the electron will choose will be the one with the least integral. There will be similar paths with low action, so it is impossible to predict exactly which path will be chosen. These integrals are complex numbers, the measure of the probability amplitude for the y to x transition.

This probability amplitude, a complex number, for the path of least action has a linear size, p, and a circular rotation, α; the real probability of the transition being p^2. This combination of linear size and circular motion in a complex number is used throughout the sciences, and is featured in the Principle.

> "A movement in a straight line cannot be sustained forever. For anything to have an eternal nature it must move in a circle."[2]

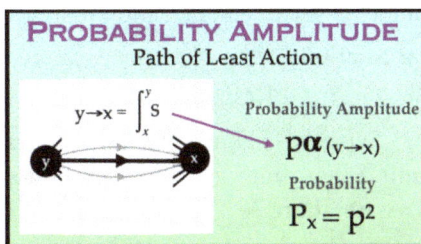

PROBABILITY AMPLITUDE
Path of Least Action
$$y \to x = \int_x^y S$$
Probability Amplitude
$$p\alpha \ (y \to x)$$
Probability
$$P_x = p^2$$

Nature involves both linear and circular changes, so it makes sense that the numbers used to describe nature need to combine both. Such numbers are called *complex* numbers while the familiar numbers that deal with linear aspects are called *real* numbers. The real numbers lie on the real axis

---

[1] *Exposition of the Divine Principle*, 1996, p. 35

[2] *Exposition of the Divine Principle*, 1996, p. 43

that stretches from minus infinity through zero to plus infinity. The operation, "rotate by 90°" is symbolized by the letter "**i**", and the real axis rotated by 90° is called the imaginary axis, stretching up and down from $-\infty i \rightarrow 0 \rightarrow +\infty i$, their only point in common being zero. These two axes define the complex plane which stretches to infinity in all directions.

For describing the probability amplitude all we need is a tiny patch of the complex plane, the unit circle with a radius of 1 centered on zero. This limit is imposed by a probability, p^2, that can never be greater than 1, a certainty, or smaller than zero, forbidden. The real axis goes from –1 through zero to +1 while the imaginary axis goes from –i through zero to +i. Just as **x** and **y** are traditionally used for real numbers, **z** is used to signify a number on the complex plane.

There are two basic ways to measure z, the rectangular and the polar. The rectangular describes z in terms of its real an imaginary components.

$$z = x + yi$$

This form makes the addition of complex numbers simple: add the x components, add the y components—$(x+ yi) + (2x+2y)= 3x+3yi$. In this form, the absolute square extending as probability from the probability amplitude is:

$$|z|^2 = x^2 + y^2.$$

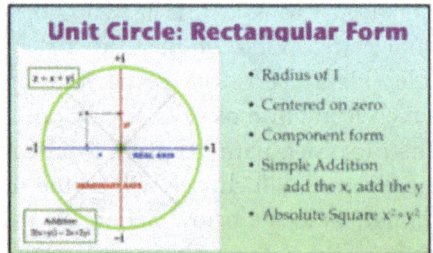

Unit Circle: Rectangular Form
- Radius of 1
- Centered on zero
- Component form
- Simple Addition
  add the x, add the y
- Absolute Square $x^2 + y^2$

The polar form views z as a little arrow with a length, p, and a rotation from the positive real axis by an angle, **α**—a linear size and an angular rotation.

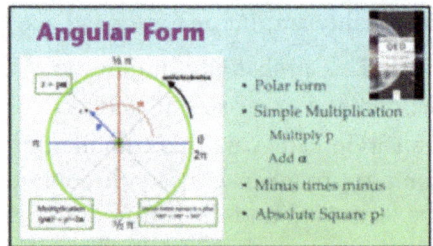

Angular Form
- Polar form
- Simple Multiplication
  Multiply p
  Add α
- Minus times minus
- Absolute Square p^2

$$z = p@\alpha \text{ (technically, } pe^{i\alpha})$$

Scientists always measure angles in radians—the distance around the unit circle—but for the nonprofessional reader we will use the familiar degree notation, $90°=\frac{1}{2}\pi$ radians, $180°=\pi$, $360°=2\pi$.

This form makes multiplying complex numbers together very simple: multiply the lengths, add the angles. In this form the absolute square is $|z|^2 = p^2$. It is this adding angles in multiplication that provides a simple reason to the schoolyard ditty:

Minus times minus is a plus,
for reasons we will not discus.

Minus one as a complex number is 1@180°, and squaring it gives *plus one*:

$$(-1)^2 = (1@180°)^2 = 1^2 @(180°+180°) = 1 @ 0° = +1$$

So the reason minus times minus is a plus is: 180° plus 180° is 360°. Note that i is the complex number 1@90° and squaring it gives 1@180°, which is the explanation why the rotation operator, i, is familiarly known as "the square root of minus-one."

There is an excellent book—*QED: The Strange Theory of Light and Matter*—by the Nobel Laureate Richard Feynman that explains much about the probability amplitude. He uses layman terms, and avoids dealing with complex numbers by "adding little arrows" and "shrink-and-turn" for multiplication (p is always 1 or less for probabilities, so multiplying them is usually a shrink).

A note to conclude this detour into complex numbers and the probability amplitude. In classical science, to calculate the probability of an **OR** situation you add the separate probabilities, while for an **AND** situation you multiply the separate probabilities, to get the final probability.

In quantum science, in **OR** situations you add the separate probability amplitudes, while in **AND** situations you multiply the separate probability amplitudes, to get the final probability amplitude. The absolute square of the result gives the final probability.

It is the subtle difference between the math of real numbers and the math of complex numbers that underlies the apparent weirdness of the quantum world.

The Wavefunction

Consider an electron, for example, constrained within a sphere such that the probability of finding the electron there is 100%. There will be a probability amplitude to move from any location within the circle to any other, including one for staying in the same place.

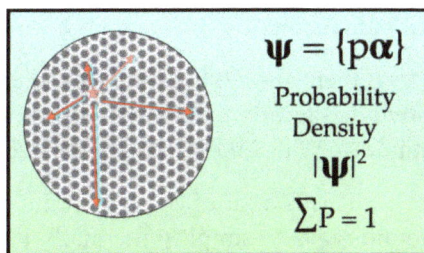

$\psi = \{p\alpha\}$

Probability Density

$|\psi|^2$

$\sum P = 1$

The set of all the probability amplitudes is called the wavefunction, usually symbolized with the Greek letter psi, **ψ**. The electron will not have the same probability of being found at all locations.

It is one thing to have a probability projected from the abstract realm, but another to know how it influences the real world. Mathematics proves that the impact on the physical realm is governed by the Law of Large Numbers (LLN). The LLN states that, given a sufficient number of tries, an abstract probability will become the real result—the more the events, the closer it will express the probability.

One example of this law in action is the Las Vegas casinos. Each game is designed so that the casino has a small probability of winning. The house edge on blackjack is 0.5%—one of the lowest—so you're looking at an average loss of 50 cents every time you bet $100. While some will win big, and some will lose big, the house is sure to make its 0.5% overall when thousands of people play.[1]

While a coin has a 50% chance of heads or tails, a single coin toss is always 100% one or the other. Tossing 100 coins almost 50% will be one or the other. Analysis shows that the error is usually close to the square root of the number of attempts. Tossing 100 coins the error would be plus or minus 10, a 10% error. Tossing 1 million coins, the results would deviate from the probability by ±1,000 coins, a percentage of 0.1%. A trillion coins, the deviation would be ±1,000,000, a 0.000,001% error. The electron moves a trillion time a second, so the difference between the abstract probability and the actual density is ±0.000,001%.

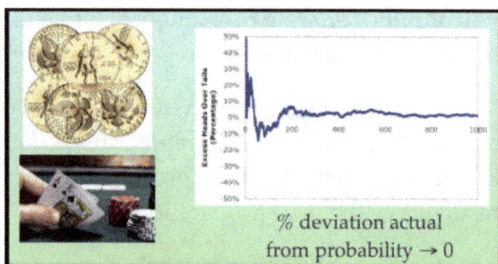

% deviation actual
from probability → 0

The topography of probability within the sphere is the absolute square of the wavefunction, and is called the probability density of the electron in this wavefunction.

If there are 100 locations, each has associated 100 tendencies to move to any other location. This wavefunction is thus a set of 10,000 probability amplitudes. With 1,000 locations; the set is a million, etc.

This is a lot of complex numbers to deal with! Luckily, scientists have found ways to simplify the math. For instance, in a helium atom the wavefunction of the nucleus and the wavefunction of the two electrons blend together to form a sphere called a 1s orbital. This sphere has a volume about a trillion times that of the nucleus.

The nucleus is about 8,000 times more massive

1s orbital
Helium atom

Combining
Probability Amplitudes
$p\alpha \rightarrow p^2$
$2\,p\alpha \rightarrow 4\,p^2$
$p\alpha + p\bar{\alpha} = 0$

Second derivative with respect to X

Schrodinger Wave Function

$$\frac{\partial^2 \psi}{\partial x^2} + \frac{8\pi^2 m}{h^2}(E - V)\psi = 0$$

Position   Energy   Potential Energy

---

[1] https://www.gamblingsites.com/blog/the-house-edge-in-blackjack-28277/

than an electron, and just sits quivering at the very center of the orbital. The lightweight electrons, however, zip about the orbital so rapidly that it appears to be solid. *Solid* is a concept built on the probability density of electrons in atoms. The actual external electron density is a full expression of the abstract probability density, itself a projection of the abstract wavefunction. A helium atom, in normal situations, has a zero probability of losing or gaining an electron, it is chemically inert.

It was Nobel laureate Erwin Schrodinger who discovered an equation that simplified calculating atomic wavefunctions. It is a differential equation involving kinetic and potential energy and can be completely solved for the hydrogen atom. For more complicated atoms, solutions are only approximations, but good enough for most situations.

The stability of the paired electrons is an example of the quantum math just mentioned. Two equal probability amplitudes constructively combining create a probability four times greater that of a single one. Two probability amplitudes destructively combining create a probability of exactly zero, it will never happen. It is this that underlies the structure of the Periodic Table of the Elements and the Pauli Exclusion Principle.

In the Principle, this internal aspect giving form to the external is called the Inner Quality and the external aspect the Outer Form:

Every entity possesses both outer form and an inner quality. The visible outer form resembles the inner quality. The inner quality, though invisible, possesses a certain structure which is manifested visibly in the particular outer form. The inner quality is called internal nature and the outer form or shape is called external form.[1]

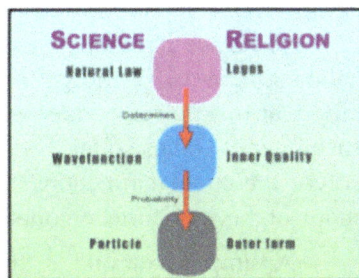

Clearly, what the Principle calls the Inner Form is what science calls the Wavefunction. In the previous talk we discussed this relation of Logos/law, the internal wavefunction, and the external form in more detail.

Chemical Activity

While a helium atom, with its highly probable electron-pair is remarkably self contained, the hydrogen atom with its lone electron, with its probability density just ¼ that of the helium electrons, is not so confined within the 1s orbital. Depending on the environment, there is a probability that another

---

[1] Exposition of the Divine Principle, 1996, p. 31

electron will cross over from elsewhere (such as a sodium atom) and create a stable pair (sodium hydride), and an even stronger probability that the electron will depart and join a chorine atom and create a stable pair there:

$$H + Na \rightarrow H^- + Na^+ \qquad H + Cl \rightarrow H^+ + Cl^-$$

A simpler stability occurs when two hydrogen atoms meet, their 1s orbitals merge into a molecular orbital, and the two electron pair-bond happily, tying the two atoms into a hydrogen molecule. This pair bond appears throughout chemistry, it is the single line in molecular diagrams, such as methane.

A more complicated situation arises in benzene, where the six simple bonds are complemented by three pairs that encompass all six carbons, making benzene remarkably stable. Such "delocalized bonds" play a significant role in adenine which plays important biological functions in energy transfer, hydrogen transfer and as a logical bit in the digital manipulations of RNA and DNA.

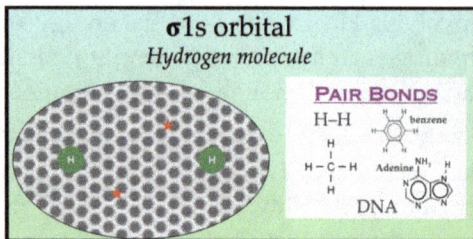

σ1s orbital
Hydrogen molecule

PAIR BONDS
H–H
H–C–H
DNA

Entangled Wavefunction

So far we have discussed simple wavefunctions, such as spheres and oblate spheroids. We now move on to more sophisticated waveforms. You might, at this point, wonder why the name is wavefunction as we have not encountered waves so far. Yet another expression for a complex number involves the circular functions, the sine and cosine of trigonometry, and the shape of the wavefunction often reflects these functions.

A simple example is the sine wave. A simple wave, such as the 1s orbital is just one half of a sine wave, with a zero at the boundary and maximum at the center, around the nucleus. The technical name for a zero in a wave is a node.

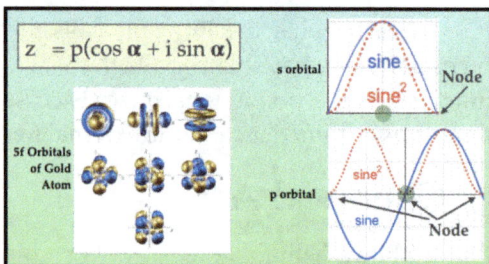

$$z = p(\cos \alpha + i \sin \alpha)$$

5f Orbitals of Gold Atom

s orbital

p orbital

sine
sine²
Node
sine²
sine
Node

The 2p orbital—there is no 1p—is a full sine wave, with nodes at either boundary, and an internal node at the very center, at the nucleus. An electron in the 2p spends 50% of its time on one side, 50% of its time in the other, and no time whatsoever at the node, at the nucleus. This node is not a barrier, however, and the electron ignores it as it flits from one side to the other filling out the shape of the orbital with a probability density.

These nodes are an introduction to an area of physics that is currently exploding—the phenomenon of entanglement.

Albert Einstein colorfully dismissed quantum entanglement—the ability of separated objects to share a condition or state—as "spooky action at a distance." Over the past few decades, however, physicists have demonstrated the reality of spooky action over ever greater distances—even from Earth to a satellite in space.[1]

Einstein considered this node aspect of quantum science a sign that something was wrong with the theory as there is no limitation on the size of a node, on the spatial separation of the zero probability between the lobes of positive probability.

In an atom, the separation between lobes is on the scale of nanometers. In the slit experiment where is seem the particle passes through both slits at the same time, it is actually two lobes of the wavefunction that pass through both slits, the particle flitting between the two as always. The lobes interfere with each other on the far side creating a pattern of probability at the detector. The separation between lobes is on the scale of millimeters.

Experiments have been performed using the sewer pipes of Vienna as protective conduits for optical fibers—scale of miles—and earth/satellite communication—scale of hundreds of miles. Theoretically, the scale is unlimited—Earth to Sirius, scale light years; Earth to Andromeda, scale millions of light years—are all possible separations for entangled lobes of a wavefunction. It is no wonder that this opens up a whole new field of technological possibilities.[2] An aspect of this is explored later in the discussion.

Wavefunction in Life

Moving into the realm of living organisms, without a doubt, the most important foundation for life on earth is the trapping of the energy in sunlight, and its storage in the form of carbohydrates. While some simple forms of life find other ways, all familiar organisms are dependent on this source of nour-

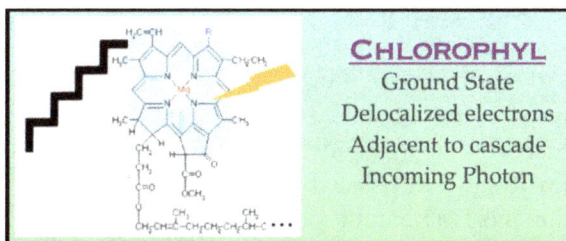

CHLOROPHYLL
Ground State
Delocalized electrons
Adjacent to cascade
Incoming Photon

[1] https://www.sciencemag.org/news/2018/04/einstein-s-spooky-action-distance-spotted-objects-almost-big-enough-see

[2] Louisa Gilder, *The Age of Entanglement: When Quantum Physics Was Reborn*, Vantage books, 2008

ishment. We can only summarize the sophisticated details here.[1]

The molecule central to this miracle is chlorophyl, and its core action involves its wavefunction and an electron donated by the magnesium ion held captive at the center.

In the ground-state, the wavefunction that this electron moves in is delocalized around the conjugated bonds of the core molecule. The long chain serves to anchor it in a membrane adjacent to an electron transport chain, a cascade of molecules, such as cytochromes, that give-and-receive electrons readily.

An incoming photon of light is absorbed by the delocalized electron and it enters an excited state orbital. In an atom, the ground-state orbital is smaller than the excited-state orbitals. In chlorophyl, it is the opposite; the ground-state orbital is delocalized while the excited-state orbital is localized at a spot adjacent to the electron transport chain.

The excited electron drops onto the electron transport chain, then as it cascades down its energy is used to generate ATP. At the end of the chain, the electron is returned to the chlorophyl in the ground-state. This is Photo System 2.

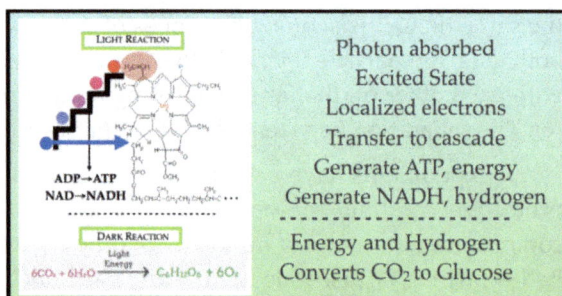

Photon absorbed
Excited State
Localized electrons
Transfer to cascade
Generate ATP, energy
Generate NADH, hydrogen
- - - - - - - - - - - - - - - - - - - -
Energy and Hydrogen
Converts CO_2 to Glucose

Alternatively, the excited electron can be channeled into Photo System 1 where it gets an extra boost and ends up generating NADH, an activated form of hydrogen. The electron-deficient chlorophyl in PS 1 gets back to the ground-state by taking an electron from a water molecule and releasing oxygen.

The ATP and NADH generated in this Light Reaction are used in the Dark Reaction to synthesize carbohydrate from carbon dioxide in a cycle of transformations. The first step in this process involves what is listed as the most abundant protein on earth.

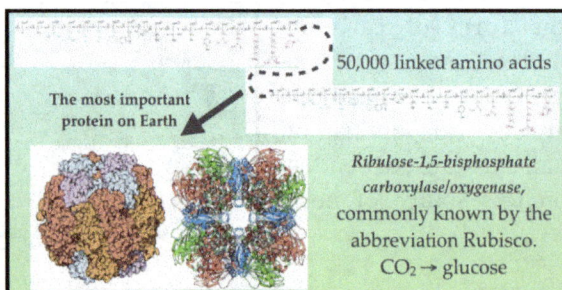

50,000 linked amino acids

The most important protein on Earth

Ribulose-1,5-bisphosphate carboxylase/oxygenase, commonly known by the abbreviation Rubisco.
$CO_2 \rightarrow$ glucose

---

[1] https://www.ncbi.nlm.nih.gov/books/NBK9861/

This protein, called Rubisco, takes a single carbon molecule (1-C) of carbon dioxide molecule, which is stable and unreactive, adds it to a 5-C molecule, and releases two 3-C molecules. This is the fixation of carbon. The ATP and NADH are used to drive the 3-C molecules back to the 5-C starting point as well as liberating the basic building blocks of carbohydrates. Rubisco is so abundant as it is rather inefficient and often mistakes an oxygen for a carbon dioxide.

PROTEIN FOLDING

Fascinating, but the key point here is the transformation of the linear chain of 500 or so amino-acids as spooled out of a ribosome into the precisely folded active form of the enzyme. This is still a major field of study with, as yet, no clear answers:

"Proteins are the workhorses of life, mediating almost all biological events in every life form. Scientists know how proteins are structured, but folding— how they are built—still holds many mysteries."[1]

The problem is that the folding takes place quite rapidly into the form that allows it to perform carbon fixation. The mystery arises because there are zillions of ways the chain could fold, and the probability of it picking the correct one is essentially zero. This, of course, is classical thinking along the lines of the Traveling Salesman Problem[2] with 500 locations involved. The Rubisco folding, in classical thinking, should take decades, not fractions of a second.

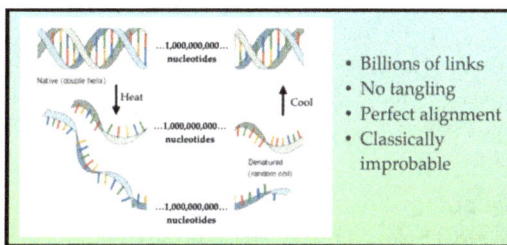

- Billions of links
- No tangling
- Perfect alignment
- Classically improbable

Once we consider the wavefunction aspect of the 500 long chain, we see that the chain is moving from an excited state to the ground-state, and the particles are just moving from an improbable state to a highly probable state.

A similar challenge involving linear chains of billions molecules is the phenomenon of DNA annealing. The iconic double helix of a DNA molecule can be disrupted by gentle heat, separating the double helix into two independent strands. The DNA is denatured. If the mixture is cooled, the two strands will align and recreate the double helix, the nucleotides hooking up with their partner in correct alignment. This is routine in the DNA poly-

---

[1] https://www.sciencedaily.com/releases/2016/03/160331134308.htm

[2] https://en.wikipedia.org/wiki/Travelling_salesman_problem

merase chain reaction that forensic science is enamored with as it allows one molecule of DNA to be multiplied into the thousands need for analysis.

This renaturing of DNA is even more classically impossible than protein folding as billions of nucleotides have to align correctly. It should take centuries, not the seconds that it actually takes. Yet when considered as a change from an excited state of the wavefunction to the ground state, it does make sense.

The latest advance in the folding problem involves computer AI learning:

Scientists from Google DeepMind have been awarded a $3 million prize for developing an artificial intelligence (AI) system that has predicted how nearly every known protein folds into its 3D shape.[1]

One would think that wavefunction-thinking would have permeated science, applied to the workings of cells all the way up to the brain, but that is for the future—biology has yet to respond to the quantum revolution in physics. My PhD thesis choice was: The Impact of the Quantum Revolution on the Biological Sciences, and after months of study, I had to report to my advisor that the impact was negligible other than in biochemistry. The response, "That was interesting in itself" was sufficient to allow me to finish my thesis.

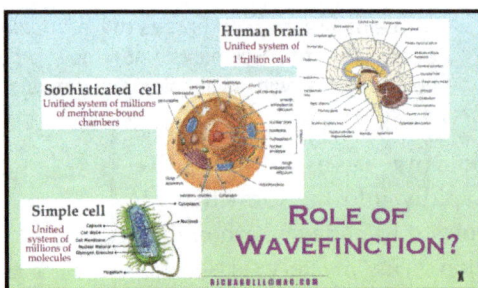

A bacterium is a unified system of millions of molecules; an animal cell is a unified system of millions of bacterial-sized, membrane-bounded chambers; the human brain is a unified system of a 100 billion neurons (a protein-like role) and 900 billion glia cells (a DNA-like role).

The concept of an all-embracing wavefunction creating all that unification has yet to enter into the scientific discussion. That is the task of future generations of scientists. For now, all we shall examine one more topic, that of the structure of protoplasm that is found in all cells.

Cytoplasmic harmony

While the nucleic acids are great at manipulating digital information to govern the state of a cell, it is the proteins who do all the actual work. In the early days of biochemistry, the cytoplasm was considered to be saline

[1] https://www.livescience.com/alphafold-wins-breakthrough-prize?utm_campaign=368B3745-DDE0-4A69-A2E8-62503D85375D

solution with the myriad large and small molecules moving randomly about in it in thermal motion. All the thousands of substrates and all the thousands of matching enzymes were randomly bumping around chaotically until they happened to encounter each and do their thing.

The technical term for this method of movement is the The Drunkard's Walk[1] which measures statistically how long it takes to get from one place to another. This gives credence to the intuition that the probability of two such random walks intersecting—assignation of enzyme and substrate—is rather small. Even smaller when the thousands of assignations that occur each second are also considered.

There are two ideas we shall examine of ways in which the cytoplasm could be ordered to make the myriad assignations probable, not improbable: 1) The cell symphony view and 2) The wavefunction view.

WATER MUSIC

While the foundational interaction of life is the covalent bonds of carbon compounds, the existential interaction is the hydrogen bonds that structure the contents of the cytoplasm. Water is structured by H-bonds between

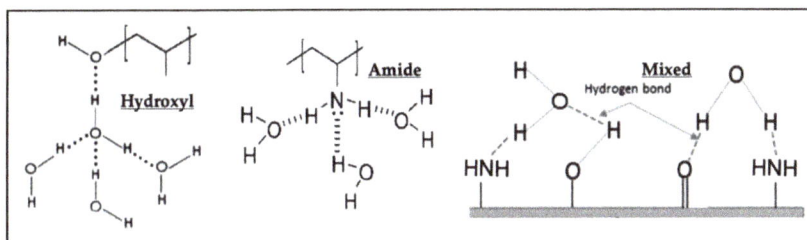

hydroxyl groups (OH), while proteins are internally structured, and structure the enveloping water with hydroxyl (OH) and amide (NH_2) radicals—which can give and receive H—and the carbonyl group that can only receive.

The all-important pair bonding of the bases in DNA and RNA is a reflection of their complementary H-bonding ability

The primary framework of the cytoplasm is its ~85% water molecules. The active portion are the many protein enzymes floating in the cytosol water.

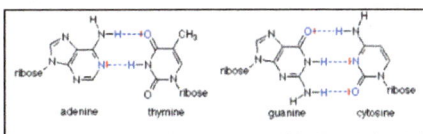

All proteins have the same basic structure: a lipid dominated core that excludes water. This is enveloped by a surface speckled with many H-donors and H-receivers.

---

[1] https://medium.com/swlh/the-mathematics-behind-a-drunkards-walk-bf41001795f2

This array, different for every type of protein, influences the dynamics of water structure for a large distance around the protein. This can be considered as analogous to a musical instrument or singer structuring the air that surrounds it.   Each biological molecule has a different, unique patter of H-bonding capacity. Like the unique sounds of sax, oboe and violin, each molecule has its unique patter of H-bond resonance in water.

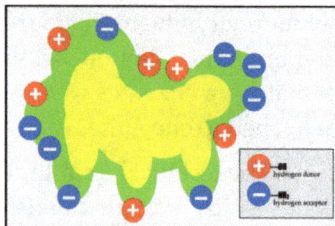

A familiar example of the capacity of proteins to organize water is Jello® where one gram of protein can structure 500 grams of flavored water. The molecular weight of gelatin is ~300,000 and that of water is 18. In solid jello, for each molecule of protein present, there are ~8,000,000 water molecules, all of them being provided a wavefunction in which they can settle into the stable ice-like structure that water is most comfortable in[1]. Jello does this remarkable feat at room temperature, not at the freezing point.

Another biological, if less appetizing, example is to watch a drop of blood flow from a small wound and then, in seconds, turn into a jello-like clot, triggered by a small amount of inactive protein in the fluid blood flipping to a gelatin-like active form and sealing the breach.

The 'music' of the cytoplasm is not in air pressure vibrations, rather it is in H-bond vibrations, and, just as in a symphony, there are emitters (aka instruments and choir) and responders (the ears of the audience and microphones of recorders).

The length of a hydrogen bond of the same type can vary over a broad area, and its spectroscopic parameters in vibrational spectra vary so much that reliable assignment of the bands in the spectrum to hydrogen bond vibrations is still a difficult problem.[2]

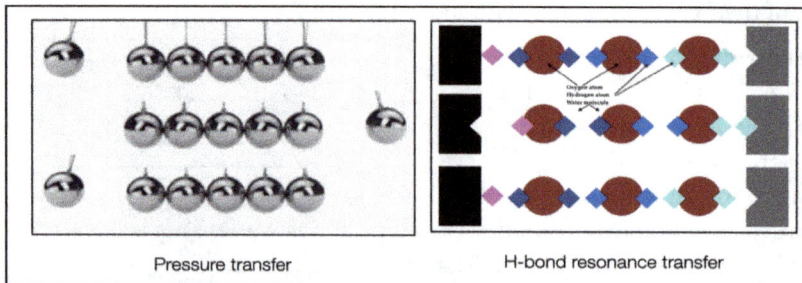

| Pressure transfer | H-bond resonance transfer |

---

[1] Technically, has least free energy.

[2] Kolesov BA. Hydrogen Bonds: Raman Spectroscopic Study. Int J Mol Sci. 2021;22(10)

If a water molecule donates in a H-bond, it is replaced by one from another water molecule, creating a 'proton wire' so to speak. "The rapid translocation of H+ along a chain of hydrogen-bonded water molecules, or proton wire, is thought to be an important mechanism…"[1]

As the cytosol has many different proteins in it, this can be considered as analogous to an orchestra playing a symphony, such as Beethoven's 9th, *Ode to Joy*. The choice of what score the cytosol is playing, of course, is decided by an output from the nucleus.

In this analogy, the audience is equivalent to the many substrates who are respond to the music. Just as a human listener has no problem distinguishing the violins, the trumpets, the harps, the piano, etc., each substrate is tuned to respond to the 'notes' emitted by their enzyme, and is attracted to it from a distance.

That is one view of an ordered cytoplasm, the second view is similar.

WAVEFUNCTION CONCERT

We noted earlier, that in an interaction of two entities, the wavefunctions of both merge into a resultant form depending on the Logos.

This implies that, at least for the outer malleable levels, the internal wavefunction spreads along the chains of H-bonded molecules which respond externally in an ordered way. We do not need to go far as to speculate about the entire cytoplasm being in an entangled state, but there is an external ordering that occurs due to the wavefunction aspect.

As noted in other sections, the internal wavefunction is not governed by spatial considerations, so if the composite wavefunction has a high probability, say, of an ATP molecule in a location lacking one, an ATP molecule can jump to fill the vacancy.

The internal wavefunction is more extensive than the external subsystems. This implies that the cross section for an interaction is much larger than the physical size. This could be experimentally confirmed in the kinetics of ultra dilute reactions, comparing predicted random walk times with actual times.

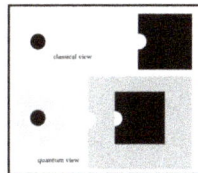

The wavefunction creates a high probability of the substrate moving towards the enzyme and a low probability of it moving away.

---

[1] https://www.ncbi.nlm.nih.gov/pmc/articles/PMC1233454/

While the wavefunction extends along the water chains, there are practical limits to how far the wavefunction and the H-bond wires can extend their influence, however, and this limit seems to be over bacteria-sized volumes. The Jello example shows, on a large scale, that this is a quite credible assumption on a small scale of ~10 billion water molecules.

These prokaryotes range roughly from 0.4 to 3 x 10^{-15} liters. Supporting this view, the much larger eukaryote cell is actually divided up into bacteria-sized compartments by internal membranes, structural tubules, etc. Each compartment interacting with the others around it. In this view of cytoplasm as bacterial symphony, then our eukaryote cells are akin to a festival of symphonies, such as Caramoor in NY, but all the orchestras playing simultaneously in different locations.

A better analogy (but not musical) is Manhattan with its many districts—garment, theater, government, law enforcement, flower wholesalers, etc.—all working and interacting together.

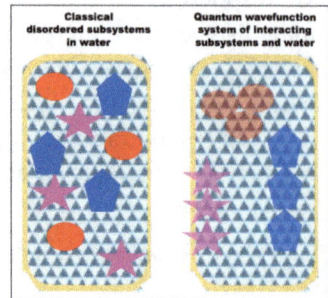

This concept of a path of probability generated by natural law acting on the wavefunction has a history. If, in the following quote, you replace the vague ideas of pathways, valleys and hills with the concept of quantum probability, high or low, it fits here just fine:

"The epigenetic landscape was a completely original contribution [by Conrad Hal Waddington] to developmental biology, [and] which was applied to evolutionary theory. The latter is a multidimensional state space in which the axes are all the attributes of the organism and the surface represents the fitness [and] there would be local 'peaks' of fitness from which organisms could not easily reach higher peaks, as they would have to get there through a 'valley' of lower fitness."[1]

It is clear to all physicists that the foundations of all modern science has been radically altered from those of the classical era. But biologists, like the cartoon character who is unaware that the ground has shifted, but will any moment—have not dealt well with this foundational change:

This is indeed the argument of many scientists who accept that quantum mechanics must, at a deep level, be involved in biology; but they insist that its role is trivial. What they mean by this is that since the rules of quantum mechanics govern the behavior of atoms, and biology ultimately involves the interaction of atoms, then the rules of the quantum world must also operate

[1] Slack, Jonathan M. W. "*Conrad Hal Waddington: the last Renaissance biologist?*" Nature Reviews Genetics. vol. 3, no. 11, Nov. 2002, pp. 889+.

at the tiniest scales within biology—but only at those scales, with the result that they will have little or no effect on the scaled-up processes important to life.[1]

While water has been a subject of scientific study for centuries now, it seems that water still has mysteries yet uncovered. A recent paper—*Topological nature of the liquid-liquid phase transition in [water]*[2]—discusses the topological structures in water, including entangled rings, high density and low density localities. Now, density variations are the very underpinnings of music, so this development suggests that 'the music in the cell' is not an unsupported conjecture.

METABOLIC SYMPHONY

Abandoning the concept of cytoplasm being a random jumble of molecules, we have in its place the concept of metabolism being akin to that of an orchestra playing a symphony. The roles in both 'organizations' are similar and the result harmonious.

Symphony	Metabolism
Digital info	score	mRNA
Waves in	air pressure	H-bond water org
Generators	singer, violins etc.	proteins, carbs etc.
Responders	audience	substrates
Organizing Center	conductor	nucleus

Rather than the chaotic mingling of the classical view, in this view we have a **meta**bolic sym**phony,** a **metaphony** in the interior of bacterial-sized volume of cytoplasm. In the eukaryote cell, our type of cell, the metaphony is a symphony of symphonies, which has no real cultural image to compare with.

---

[1] McFadden, Johnjoe; Al-Khalili, Jim. *Life on the Edge: The Coming of Age of Quantum Biology* (p. 19). Crown/Archetype.

[2] Andreas Neophytou et al, (2022) Nature Physics https://doi.org/10.1038/s41567-022-01698-6

4 • A UNIVERSE
JUST RIGHT FOR LIFE

As modern science has delved deeply into the structure of reality, it has become apparent that we inhabit a universe in which many parameters and laws are finely adjusted to allow for the presence of life. These have been extensively documented in books by scientific authors.

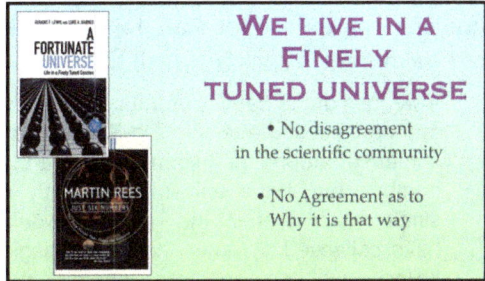

WE LIVE IN A FINELY TUNED UNIVERSE

• No disagreement in the scientific community

• No Agreement as to Why it is that way

While this is not in dispute, there is no agreement as to what caused this fine-tuning. One book on the topic concluded with a discussion of the various theories:

> Our conclusion is that the fundamental properties of the universe appear to be fine-tuned for life.... We would like to know: why is the universe like this?... The ideas inspired by the fine-tuning of the universe for life range from realistic science to informed guesswork to unfettered speculation.[1]

In this book we will just discuss two contrasting possibilities: the universe is designed for life by a creator God, or, the universe is a random serendipity, one of a multiverse that happens to be just right for life.

While it is quite possible to sense the hand of God in the stars, the flowers, the people, etc., these are topics for debate so we shall dig much deeper into the very foundations of all the wondrous aspects of our reality.

First, we will examine what fine-tuning is to be found at a fundamental level; secondly, examine the two very different ways of explaining it all.

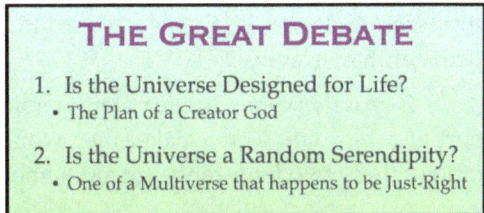

THE GREAT DEBATE

1. Is the Universe Designed for Life?
   • The Plan of a Creator God

2. Is the Universe a Random Serendipity?
   • One of a Multiverse that happens to be Just-Right

---

[1] G. F. Lewis, L. A. Barnes (2016) *A fortunate universe: Life in a finally tuned universe*, Cambridge University Press, p. 290

The Big-Bang

We will start at the very beginning, with the Big-Bang and a very important ratio of the actual density of the universe and the critical density.[1] For a flat universe such as ours today the ratio is approximately unity.

As the universe expands, any deviation from unity is magnified. A small deviation at the start to less than one results in runaway inflation into a universe too diffuse for stars to form. A small deviation to greater than one results in a big crunch in which the universe collapses quite rapidly.

If we look at the density of the universe just one nanosecond after the Big-Bang it was immense, around 10^{24} kg/m³. This is a big number, but if the universe was only a single kg/m³ higher, the universe would have collapsed by now. And with a single kg/m³ less the universe would have expanded too rapidly to form stars and galaxies.[2]

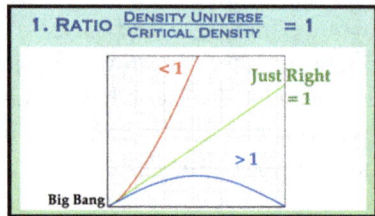

1. RATIO DENSITY UNIVERSE / CRITICAL DENSITY = 1

< 1   Just Right = 1   > 1   Big Bang

For neither of these catastrophes to happen, the ratio at the very beginning of the Big-Bang had to be exactly one to an accuracy of 10^{60}. The Big-Bang was tuned to an extraordinary degree to allow our universe to be just right for life.

Two long-range forces emerged from the Big-Bang: the force of gravity and the force of electromagnetism. Gravity is by far the weakest of the forces, but it has the advantage that all matter has positive gravitation. Electromagnetism (EM) is by far the stronger but is usually neutralized

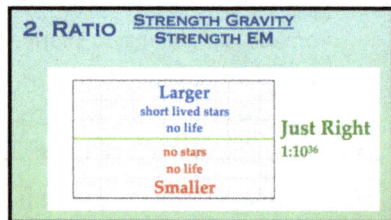

2. RATIO STRENGTH GRAVITY / STRENGTH EM

Larger
short lived stars
no life
no stars
no life
Smaller

Just Right
1:10³⁶

by positive and negative charges in equal amounts. The ratio of their strengths is $1:10^{32}$ — EM is an astounding billion, trillion, trillion times stronger than gravity.

If gravity were stronger, stars would be short lived and unable to support life; if gravity were weaker they would be no stars at all, and no life. The ratio of these two long-range forces is finely tuned to allow for life.

The next topic involves the two short-range forces that emerged from the Big-Bang, the strong and the weak nuclear forces.

---

[1] The 'critical density' is the average density of matter required for the Universe to just halt its expansion, but only after an infinite time. A Universe with the critical density is said to be flat.

[2] G. F. Lewis, L. A. Barnes (2016) *A fortunate universe: Life in a finally tuned universe*, Cambridge University Press, p. 167

The Big-Bang generated hydrogen 75% and helium 25% along with an abundance of photons and neutrinos. For most of their lifetime, stars generate their profligate energy by converting hydrogen into helium. This essential process is very sensitive to the ratio of its efficiency to the strength of the strong force, in the appropriate units. If it were a little higher, the combination of two protons would be stable, and the star would rapidly convert all its hydrogen and explode in the process. If it were a little weaker, the essential intermediate of deuterium—its neutron generated by the weak force—would be unstable and stars could not convert hydrogen at all.

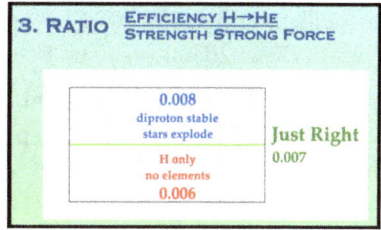

3. RATIO	EFFICIENCY H→HE STRENGTH STRONG FORCE
0.008 diproton stable stars explode	Just Right
H only no elements	0.007
0.006	

This is another ratio that has to be just right to allow for stars that can support life.

One of the most astonishing advances in modern science is that all the amazing, beautiful and complex entities in creation are constructed out of just three entities (like a Lego set with only three types of pieces).

These fundamental building blocks are the U and D quarks of the atomic nucleus, and the electrons that surrounds them in atoms. Their relative masses are significant (note that a nucleon of three quarks is ~2,000 times the mass of the electron—most of this extra mass being in the field of energy binding the quarks together). The balance is such that a neutron (DDU) can decay into a proton (DUU) and an electron.

If the electron or U-quark mass were larger, the proton would be unstable and the universe a neutrons-only wasteland. If the D were larger, the nuclear neutrons would be as unstable as isolated ones, and the universe a sterile hydrogen-only wasteland. These three entities emerging from the Big Bank are balanced just right for a fecund universe.

That we are all constructed from molecules of carbon and oxygen is only possible because the masses of the quarks and the strength of the forces lie within an outrageously narrow range![1]

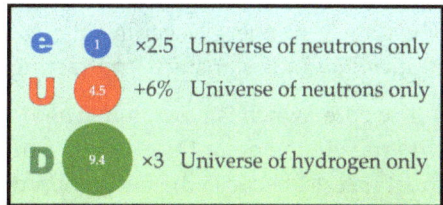

e	1	×2.5	Universe of neutrons only
U	4.5	+6%	Universe of neutrons only
D	9.4	×3	Universe of hydrogen only

Star Lifetimes

Our Sun has been a source of light and heat for the last 5 billion years —converting 600 million tons of hydrogen every second into 596 million tons

---

[1] G. F. Lewis, L. A. Barnes (2016) *A fortunate universe: Life in a finally tuned universe*, Cambridge University Press, p. 120

of helium and radiating 4 million tons of energy. It will continue to do this for another ~20 billion years before it runs out of hydrogen to convert.

The Earth, which formed at about the same time, has over four billion years evolved from a barren sphere of molten rock to the pleasant home of humanity with the plethora of animals and plants it is today. The longevity of the sun played an important role in energizing this long and complicated process. A short-lived sun would be useless.

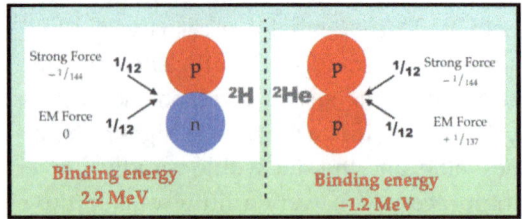

This stellar longevity depends on two parameters.
1. The weak force has to be so feeble that the transformation of protons into neutrons takes billions of years, allowing the key intermediate of heavy hydrogen to be generated slowly.
2. The strong force must not be too brawny, so allowing two protons to stick together as He-0 and causing a runaway explosion that would destroy a star. It is when two protons are briefly in contact that the weak force can flip one of them into a neutron, forming a deuteron.

The balance has to be just right: the strong force is so short range that only where the nucleons touch can it work, ~1/12th surface overlap. The EM force has no such limitation and the repulsion between two protons is sufficient to overcome the strong attraction. The nascent neutron, having no charge, allows the deuteron pair a moderate—on the nuclear level—stability of 1.2 MeV. Two deuterons are rapidly converted into very stable helium-4 with the substantial binding energy of 27 MeV.

So, as the eons pass since the Big-Bang, the relic hydrogen has been gradually converted into helium to mingle with the relic helium. A universe of just hydrogen and helium, however, is not very interesting, the rest of the elements have to be made. This happens when stars run out of hydrogen. The source of energy keeping the star inflated against gravity's pressure is reduced and the star starts contracting.

The core, which is now almost all helium, is compressed and its temperature rises to the point that Helium starts to fuse. How this happens was a great puzzle, because the natural product, beryllium-8 is utterly unstable and falls apart in a million-billionth of a second, leaving no time for a chance encounter with another helium to form carbon.

It was Dr Hoyle—the originator of the hoped-to-be derogatory term 'Big-Bang'—who figured out the answer. He reasoned that beryllium-8 must have a specific resonance that had a stability sufficient for a third helium to arrive. Furthermore, he realized the carbon-12 would also have to have a specific resonance that gave it stability. This is the triple alpha process. Finally, he realized that oxygen must *not* have a suitable resonance, otherwise all the carbon would get turned into oxygen.

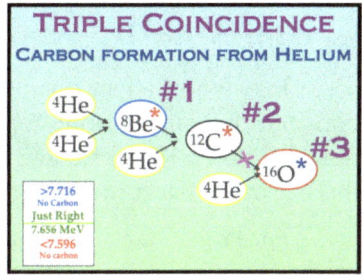

He calculated the resonance energies that were required for carbon synthesis, and all his predictions were validated when tested.

> The triple alpha process is extremely important is determining the elemental composition of the universe and allowing life as we know it to exist. Yet that the process occurs at all is somewhat improbable, as its discovery showed it was only made possible by the complex interplay of physical constants that cause the excited resonance of C-12 to occur where it does. The philosophical and scientific implications of this have prompted much discussion.[1]

This brings us to the balance the four fundamental forces crucial to the functioning of our Sun, the provider of all the energy needed to supply our Earth. Their strengths vary across 36 orders of magnitude; two are short-ranged, two are long-ranged, yet they all work in harmony to provide a perfect balance for the functioning of our Sun.

The Sun, as you might imagine, is very hot at the core, and this heat expands the Sun. Gravity, on the other hand, tends to compress the Sun, so in the stable configuration, they balance each other.

The strong force unites hydrogen into helium, releasing plenty of energy. The rate is controlled, however, by the weak force, which has to convert protons into neutrons, which takes ~9 billion years on average.

The balance between these two forces allows the Sun to remain stable in the 'main sequence' burning hydrogen slowly and steadily for billions of years.

---

[1] http://large.stanford.edu/courses/2017/ph241/udit2/

Star Death

Eventually, all stars run out of fuel. Our Sun will end up making carbon from helium, but not get hot enough to burn carbon into heavier elements. It will settle in old age as a white dwarf with the mass of the Sun but the size of the Earth cooling slowly into darkness.

For larger stars, say 20 times the sun's mass, the helium burning progresses through carbon, oxygen, neon, magnesium, sulfur and silicon fusion—generating less and less energy at each stage—until iron is formed. This is the dead-end point as no more energy can be extracted from rearranging nucleons. All the elements so crucial to life have been formed, but they are locked away in a massive star.

The production of energy in the core ends, and the gravitation contraction commences. Eventually, the core conditions become so extreme that electrons are forced to combine with protons into neutrons and neutrinos, and the core collapses into a massive nucleus with the mass

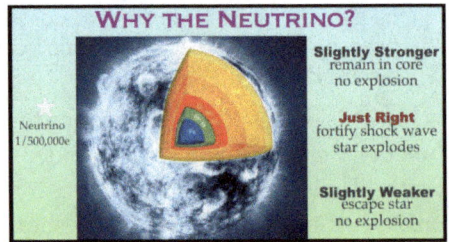

WHY THE NEUTRINO?

Neutrino
1/500,000e

Slightly Stronger
remain in core
no explosion

Just Right
fortify shock wave
star explodes

Slightly Weaker
escape star
no explosion

of the Sun and size of Mt. Everest. The loss of support at the center, causes the remaining 19-suns worth of mass to plummet inwards. The rebound is a shock wave that explodes towards the surface.

This shock wave, pushing 19-suns worth of mass outwards, would stall if it were not for the immense number of neutrinos released by the core. These energize the shock wave which then explodes the star as a supernovae that temporarily outshines an entire galaxy of 100 billion stars.

> The core collapse phase is so dense and energetic that only neutrinos are able to escape. ... The two neutrino production mechanisms convert the gravitational potential energy of the collapse into a ten-second neutrino burst, releasing [a huge energy flux] ...Through a process that is not clearly understood, ... the energy released (in the form of neutrinos) is reabsorbed by the stalled shock wave, producing the supernova explosion. Neutrinos generated by a supernova were observed in the case of Supernova 1987A, leading astrophysicists to conclude that the core collapse picture is basically correct.[1]

The life-essential elements are now freed, and dispersed for the next round of solar-system creation. Our Sun and solar system are 3rd generation, and have inherited the elements created by the 1st and 2nd generation of

---

[1] https://en.wikipedia.org/wiki/Type_II_supernova

stars. All the carbon, oxygen, etc, that our bodies rely on are there thanks to the tiny neutrino!

> "The weak interaction has to be just right to allow enough neutrinos both to escape from the core and to interact with the shock wave."[1]

We have now described (some) of the fundamental parameters that govern the universe and how they are precisely located at a Goldilocks peak that falls off rapidly on either side. The evidence of this precise and small peak of parameters suitable for life is irrefutable. The only open question is: Why?

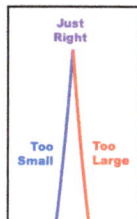

Chemical Beneficence

So far we have discussed the fine-tuned parameters of physics, and now we shall briefly venture into the realm of chemistry. The most important molecule, by far, for life is water. It has too many essential properties to list them all here.[2] Without water we have a lifeless desert, with just a little water, there is life.

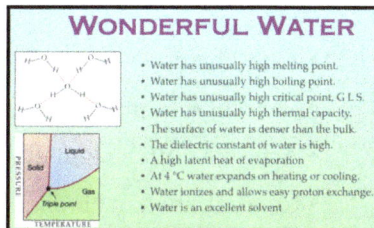

WONDERFUL WATER

> Water is the most important liquid for our existence and plays an essential role in physics, chemistry, biology and geoscience. What makes water unique is not only its importance but also the anomalous behavior of many of its macroscopic properties. The ability to form up to four hydrogen bonds (H-bonds), in addition to the non-directional interactions seen in simple liquids, leads to many unusual properties such as increased density on melting, decreased viscosity under pressure, density maximum at 4 °C, high surface tension and many more. If water would not behave in this unusual way it is most questionable if life could have developed on planet Earth.[3]

The anomalous properties of water are numerous, and many are important to life. While some are of lesser consequence, a list of up to 70 key properties have been published.[4]

Water is essential to biochemistry, the molecules of life. The driving force of water molecules seeking minimal energy hydrogen bonds, helps in the folding of linear amino-acid chains into the active folded form, and the hydration shells of proteins and DNA that help maintain the active structures.

---

[1] Gribbin, John. *Cosmic Coincidences* (2014) ReAnimus Press. p. 250

[2] http://www.lsbu.ac.uk/water/index.html

[3] https://www.ncbi.nlm.nih.gov/pmc/articles/PMC4686860/

[4] http://www1.lsbu.ac.uk/water/anmlies.html

How water relates to and interacts with those systems—like DNA, the building block of all living things—is of critical importance, … a previously unknown characteristic of water. "DNA's chiral spine of hydration," … is the first observation of a chiral water superstructure surrounding a biomolecule. In this case, the water structure follows the iconic helical structure of DNA.[1]

The addition and subtraction of water molecules in cell metabolism is ubiquitous, and for plants, water serves as a source of the hydrogen needed to convert carbon dioxide into carbohydrates, liberating the oxygen do necessary to animals. It should be noted that the properties of carbon dioxide are only secondary to water in life's requirements.

These are some of the important properties of carbon dioxide:

• Molecules are not attracted to each other, so not a solid at room temperature

• A gas that is an animal waste product and nourishment for plants

• Molecules soluble in water

• Acts as a buffer, regulating the acidity of blood

• Concentration in air and water are equal

• Critical role in maintaining the temperature of the Earth

• Complex carbon cycle in air, water, soil, and deep mantle.

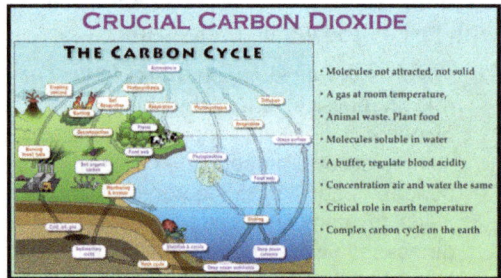

CRUCIAL CARBON DIOXIDE
THE CARBON CYCLE

• Molecules not attracted, not solid
• A gas at room temperature,
• Animal waste. Plant food
• Molecules soluble in water
• A buffer, regulate blood acidity
• Concentration air and water the same
• Critical role in earth temperature
• Complex carbon cycle on the earth

We will not deal with any of the other elements with properties essential to life—such as nitrogen, phosphorus, calcium, etc—as the point of being *Just Right* has been made.

Two Explanations

We have looked at just a few of the parameters that rule the universe. There are many, many more parameters that were established at the Big-Bang, all set so that the resultant universe was Just Right for life and human beings. There are

THE GREAT CHOICE

1. Is the Universe Designed for Life?
   • The Plan of a Creator God

2. Is the Universe a Random Serendipity?
   • So many possibilities: Multiverse needs 10^{500}

---

[1] Cornell University. "Water forms 'spine of hydration' around DNA, group finds." ScienceDaily., 25 May 2017. <www.sciencedaily.com/releases/2017/05/170525141530.htm>

thousands of settings that have to be Just Right for life—and humans—to flourish. There are a variety of responses to this fact of the universe:

Some would say—the Weak Anthropic Principle—that we are here, aren't we, so it has to be that way or we wouldn't be here to question the laws. Philosophy has dealt with this "so what" view of reality and its many, many parameter settings that are Just Right for life:

> The Canadian philosopher John Leslie has offered a neat analogy. Suppose you are facing execution by a fifty-man firing squad. The bullets are fired, and you find that all have missed their target. Had they not done so, you would not survive to ponder the matter. But, realizing you are alive, you would legitimately be perplexed and wonder why.[1]

THOUSANDS OF JUST-RIGHT FEATURES
We are here, so of course we fit the Universe!

• A fifty-man firing squad
• All Miss you!
• Surprised to be Alive!
• Naturally question 'Why?'

There are many, many parameters that have to be just right for life to exist. Just like the firing squad where just one on-target bullet would be fatal, if just one of these parameters was set wrong, there would be no life possible. So the question "Why?" is justified.

One fallacious idea is that the natural laws went through some sort of Darwinian variation until the correct ones were chanced upon. While a case can be made of biological laws, the ones we have been examining are those of fundamental significance, and are not subject to variation and selection:

> The fundamental laws and constants of nature did not gradually evolve into the present life-supporting character through a process of natural selection, as is widely believed, instead they spontaneously came into existence with the origin of the universe itself, perfectly calibrated and ready for action.[2]

Unwilling to abandon Darwin, the variation and selection has been pushed back to pre-Big-Bang times and the concept of many universes, each with a random set of laws and constants. Our universe just happens to be the one where they are all Just Right for life and for us to evolve.

This is the concept of the multiverse. As there are so many settings, and so many possibilities for each setting, there has to be a great deal of universes for even one of them to have the correct set of para-

MULTIVERSE
One
of
1,000,000,000,000,000,000,000,000,000,000,0
00,000,000,000,000,000,000,000,000,000,000,
000,000,000,000,000,000,000,000,000,000,00
0,000,000,000,000,000,000,000,000,000,000,0
00,000,000,000,000,000,000,000,000,000,000,
000,000,000,000,000,000,000,000,000,000,00
0,000,000,000,000,000,000,000,000,000,000,00
00,000,000,000,000,000,000,000,000,000,000,
000,000,000,000,000,000,000,000,000,000,00
0,000,000,000,000,000,000,000,000,000,000,0
00,000,000,000,000,000,000,000,000,000,000,
000,000,000,000,00,000,000,000,000,000,000
Others

---

1 Gribbin, John. Cosmic Coincidences (pp. 267-268). ReAnimus Press.

2 M. A. Corey, The God hypothesis: Discovering design in our 'just right' Goldilocks universe, Rowan and Littlefield, 2001, p.11

meters.

A typical suggestion is that we are one of 10^{500} universes, all of which are inherently unobservable by being not in our universe for examination. While 500 does not seem excessive, the powers of ten really add up, and the suggested number in the Multiverse it actually humongous. It is far, far larger that the 10^{99} photons of light filling the universe, the probability of winning fifty Powerball Jackpots in a row, of tossing 1,500 fair pennies and having every single one of them come up heads. A truly staggering number of universes.

The alternative is a lot simpler, and science and religion are already in partial agreement about it. Before anything substantial existed, all scientists agree that before the Big-Bang the truths of mathematics existed. This is never explicitly stated, but all the atheistic multiverse theorists use mathematics in describing their theories.

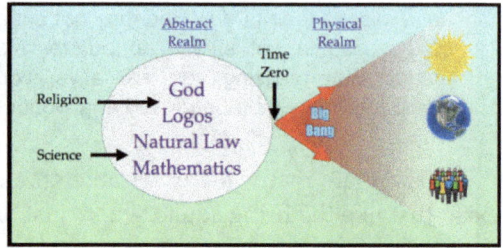

The logic of mathematical proof is these truths could not possibly be any other way. It is impossible, for instance, for the square root of two to be a fraction. There was an Abstract Realm before the Big-Bang in which mathematical truth existed. While science never explains exactly what a Natural Law is—that is for philosophers to debate—all science is based on the assumption that the world obeys these natural laws, and that it is a worthwhile endeavor to figure out what these abstract laws are.

Religion agrees that this Abstract Realm existed before the Big-Bang, but goes much further by including an intelligent, abstract Creator who designed the laws and initiated the Big-Bang.

This is the choice that science is now dealing with:

1. One well-crafted universe designed for life and the eventual emergence of humans.

2. A random assembly of zillions of universes, our one being accidentally just right for life and humans.

• One God; One Universe
• One of a multiverse of 10^{500} universes

OCCAM'S RAZOR
A Parsimonious Shave Every Time!

When faced with two opposing explanations for the same set of evidence, our minds will naturally prefer the explanation that makes the fewest assumptions.

As neither God nor the putative other universes are available for direct, scientific study, there is nothing to inform our choice except the philosophical principle: sufficient and simple explanations are better than complicated ones with a plethora of assumptions. As we survey all the evidence, the thought insistently arises that some supernatural agency must be involved. Is it possible that suddenly without intending to we have stumbled upon scientific proof of the existence of a Supreme Being?[1]

This is a controversial statement. Is it proof in the absolute mathematical sense, or in the relative jury sense of 'beyond reasonable doubt'?

One thing that is happening is that, when clearly understood, scientist's are beginning to recognize that there is an intelligence involved:

Hawking, quoted by Ian Barbour, writes, "The odds against a universe like ours emerging out of something like the Big Bang are enormous. I think there are clearly religious implications." Going even further, in A Brief History of Time, Hawking states: "It would be very difficult to explain why the universe should have begun in just this way, except as the act of a God who intended to create beings like us." Another distinguished physicist, Freeman Dyson, after reviewing this series of "numerical accidents," concludes, "The more I examine the universe and the details of its architecture, the more evidence I find that the universe in some sense must have known we were coming."[2]

1 George Greenstein (1988) *The symbiotic universe*, William Morrow p. 27

2 quoted by Collins, Francis S. *The Language of God: A Scientist Presents Evidence for Belief* (pp. 75-76). Free Press.

5 • SCIENCE IN THE REALM OF SPIRIT

For the longest time, humanity's intellectuals thought that the Earth was the main event at the center of all things; the Sun was there for illumination, the Moon to count time, and the planets and fixed stars in the furthest shell to foretell the future and look pretty at night.

This Ptolemaic view was reorganized by Copernicus who made an excellent case for the Sun being at the center, but other than that the Solar System was still paramount.

It was only in the 20th century that it became clear that the Solar System was a tiny part of a spiral galaxy of billions of suns that, to human eyes, looked like milk splashed in the heavens, hence the Milky Way. The Solar System was not at the center of the spiral—where a massive Black Hole reigns and is horribly hostile—but halfway to the periphery.

It did not take long before the astronomers found that our beloved home galaxy was just one of billions of others, and that our Local Group of galaxies

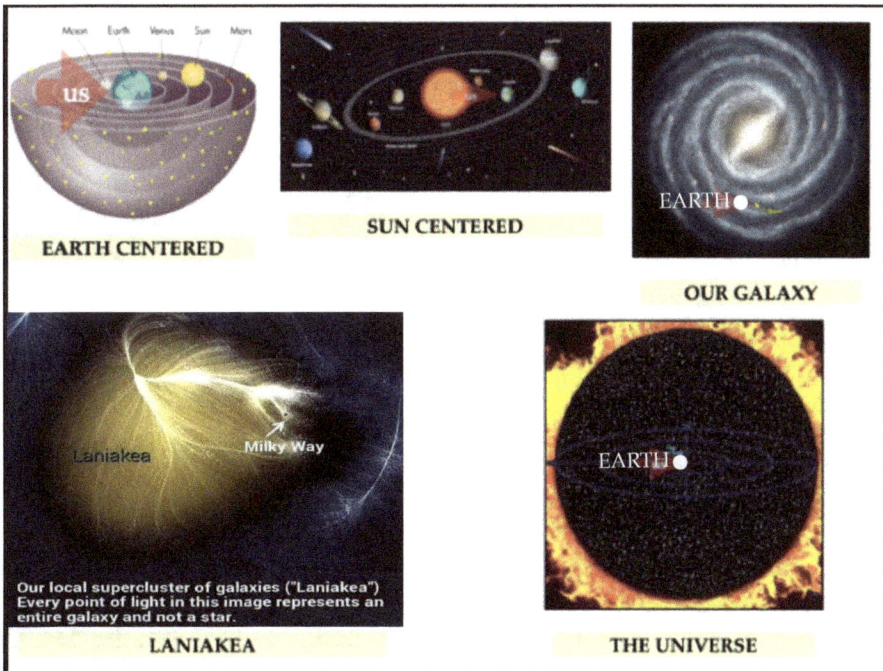

EARTH CENTERED

SUN CENTERED

EARTH●

OUR GALAXY

Laniakea   Milky Way

Our local supercluster of galaxies ("Laniakea") Every point of light in this image represents an entire galaxy and not a star.

LANIAKEA

EARTH●

THE UNIVERSE

inhabited the suburbs of the mighty Laniakea Supercluster of 100,000 galaxies.

Finally, with the discovery of the Cosmic Microwave Background (CMB), we found ourselves at the very center of the Visible Universe, a sphere with a radius of 13.5 billion light years. The CMB is the boundary, and is the wall of fire that emerged from the Big-Bang, a blaze now faded into the microwave spectrum. Ptolemy, ignoring questions of scale, was correct in placing the Earth at the center of the Universe. More correctly, putting humans at the center of the visible universe, for whatever galaxy you reside in, you will always be at the center of a 13.5 billion lightyear sphere.

Dark Matter

That was the apogee of our sense that science understood the universe. It was all about galaxies similar to ours.

The first sense that there was more going on came from the basic science of rotation. *Synchronous rotation*, exemplified by points on a vinyl record, are all rotating together as one. A*synchronous rotation*, is exemplified by the planets rotating about a central mass, the Sun, each having a different period of rotation—e.g. Mercury's 88 days, and Neptune's 165 years.

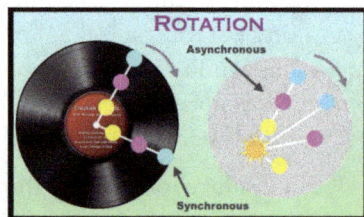

The Sun is the pivot for the asynchronous solar system, and it was expected that the Black Hole at the center of the Milky Way—with a mass of 4,000,000 Suns—would be the pivot for an asynchronous rotation of the galaxy. This expectation was incorrect, the galaxy is more synchronous than asynchronous, and without enough visible mass to hold it together.

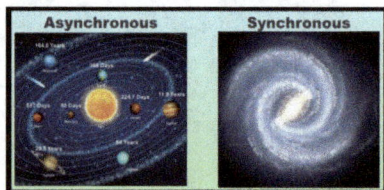

Dark matter was first hypothesized in order to account for the rotation of galaxies, which didn't seem to have enough conventional matter to keep them from flying apart like a smoothie in a lidless blender.[1]

The explanation for this, and for many other puzzlements, was that the galaxy is imbedded in a vast and massive halo of stuff, called Dark Matter—because we

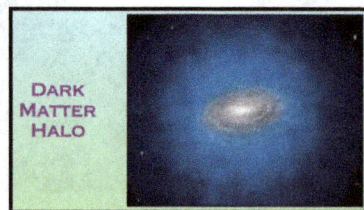

---

[1] https://www.space.com/39577-rotating-galaxy-group-raises-questions.html

cannot see it and still do not know what this stuff is—that is rotating and carrying the embedded galaxy along with it.

Measurements have shown that Dark Matter is ubiquitous, and that there is five times as much of it as there is regular matter in the universe. Science went from the hubristic sense of knowing it all to a humble admission that 80% of the universe was a known unknown.The indignity did not stop there.

Dark Energy

The huge amount of matter—regular and dark—that emerged from the hot Big-Bang had an enormous gravitational pull that was opposing the expansion of the universe. It was generally assumed, naturally, that the expansion of the universe was decelerating, it was getting slower. There was even a possibility that the expansion would stop, and that the universe would then start contracting and end up in a Big Crunch.

With this potential doomsday in mind, scientists started a series of observations to measure the rate of deceleration.

Then came 1998 and the Hubble Space Telescope (HST) observations of very distant supernovae that showed that, a long time ago, the universe was actually expanding more slowly than it is today. So the expansion of the universe has not been slowing due to gravity, as everyone thought, it has been accelerating.[1]

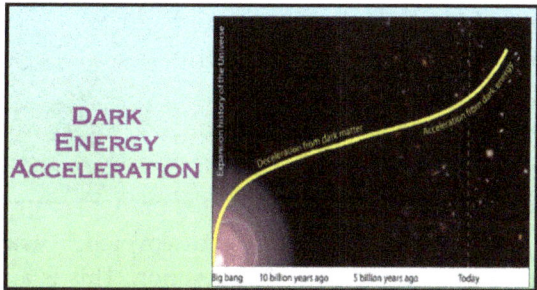

The results were unexpected: for the first 8 billion years or so, the expansion was decelerating as expected. But about 5 billion years ago—coincidentally about the time the Earth and Sun were forming—the deceleration turned into an acceleration and the expansion started speeding up.

Whatever the cause—and no one has yet come up with a convincing explantation—it was powerful enough to, not only overcome the inward gravitation of all the matter, but surpass it with an outward energetic push. This anti-gravitational vigor is called Dark Energy, and there is 70% more of this anti-gravitational energy than all the gravitating matter in the universe.

---

[1] https://science.nasa.gov/astrophysics/focus-areas/what-is-dark-energy

Another known unknown, and a further blow to science's claim of omni-science. Currently, the score[1] is:

68%	**Dark Energy**
27%	**Dark Matter**
5%	**Regular Matter**

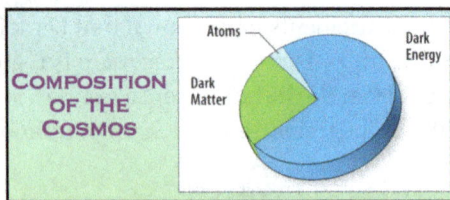

COMPOSITION OF THE COSMOS

There are many explanations circulating in the science world, we shall explore one here that has the advantage of being compatible both in the science realm and in the religious realm.

Another Realm

A clue to what this anti-gravitational energy might be was uncovered by Dr. Dirac, a quantum pioneer, in his effort to unite quantum physics with relativity.

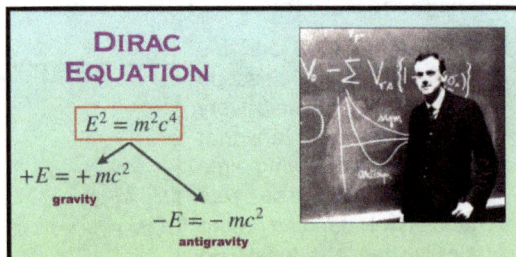

DIRAC EQUATION

$$E^2 = m^2c^4$$

$$+E = +mc^2$$ gravity

$$-E = -mc^2$$ antigravity

Dirac came up with an equation that had two solutions. One solution was the familiar Einstein relation of mass and energy. The other was similar, but introduced the concept of negative energy and negative mass.

When antimatter was discovered, it was thought that Dirac had predicted it with his negative solution. This is a mistake, however, because antimatter has as much gravitating positive mass as does regular matter. When matter and antimatter combine, they do not neutralize each other but convert into a burst of positive energy. When an electron with 0.5 MeV mass meets a positron with 0.5 MeV mass they annihilate into photons with 1 MeV of energy.

> Dirac's equation calls for both positive and negative energy... negative energy would merely be a vibration of charges at right angles to ordinary dimensions in an *imaginary* direction.[2]

Exploring the concept of negative mass-energy, theoreticians came up with the concept of Tachyons. Tachyons are distinguished from regular matter—the tardyons—by the following properties:

---

[1] https://science.nasa.gov/astrophysics/focus-areas/what-is-dark-energy

[2] https://ethw.org/w/images/7/72/PVB_Dirac%27s-sea-of-negative-energy-%28part_2%29.pdf

• Tardyons have positive mass-energy; • Tachyons have negative mass-energy.

• For tardyons, the speed of light is an upper limit which can be approached asymptotically, but never reached; • For tachyons. the speed of light is a lower limit which can be approached asymptotically, but never reached.

• Adding positive energy to a tardyon increases its speed; • Adding negative energy to a tachyon decreases its speed.

• For tardyons, the lowest speed is asymptotically Absolute Zero; • for tachyons the lowest speed is asymptotically the speed of light.

SPIRIT WORLD ENERGY

LOSS OF NEGATIVE ENERGY	High Speed, High Level, Heaven Live for the sake of others Better to Give than to Receive Sacrifice self: Lose PW life to gain SW life
GAIN OF NEGATIVE ENERGY	Low Speed, Low Level, Hell Live for the sake of self Take from others Sacrifice others: Gain PW life to lose SW life

• For tardyons, the never-to-be-reached upper limit is light speed; • for tachyons, the never-to-be-reached upper limit is infinite.

> You can now deduce many interesting properties of tachyons. For example, they accelerate… if they lose energy…. Furthermore, a zero-energy tachyon is "transcendent," or moves infinitely fast. This has profound consequences…. the problem is that we can get spontaneous creation of tachyon-anti-tachyon pairs, then do a runaway reaction, making the vacuum unstable.[1]

We will discuss this instability problem shortly. Just as the tardyons in the physical world are of different varieties, we might expect that the tachyons also have varieties. There is a powerful area of math called group theory that has great success is predicting and organizing the particle zoo of the last century. One aspect of this discipline, called supersymmetry, SUSY predicts that all the fundamental particles in the physical world, have a super symmetric partner.

The big idea of SUSY is that there could be an additional symmetry present — between fermions and bosons — that similarly protects the properties of matter and enables the particle masses to be so small …Sure, you have to double the number of known fundamental particles, creating a super-partner particle counterpart (a super-fermion for each Standard Model boson; a super-boson for each Standard

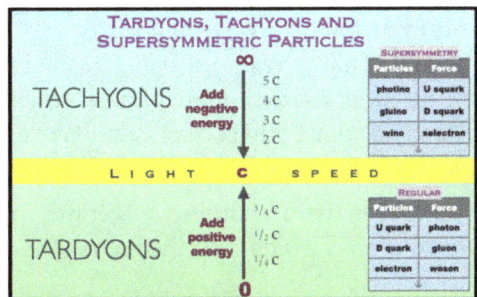

TARDYONS, TACHYONS AND SUPERSYMMETRIC PARTICLES

TACHYONS — Add negative energy

∞	
5 C	
4 C	
3 C	
2 C	

SUPERSYMMETRY

Particles	Force
photino	U squark
gluino	D squark
wino	selectron

LIGHT   C   SPEED

TARDYONS — Add positive energy

3/4 C	
1/2 C	
1/4 C	

REGULAR

Particles	Force
U quark	photon
D quark	gluon
electron	woson

0

---

[1] http://math.ucr.edu/home/baez/physics/ParticleAndNuclear/tachyons.html

Model fermion) for every one that's known.[1]

There are two problems: 1. If tachyons were to zip through our physical universe, like a speedboat on a lake, they would leave a wake of evidence that would be unmistakable, let alone them going back in time. 2. For all the efforts of experimentalists, not a single one of these mirror image particles has been found in the physical universe.

Two sides of Spacetime

If, however, tachyon super-symmetrical particles were of negative energy on the other side of space-time, this absence would make sense.

As these phenomena have not being encountered, the question is are there aspects of the cosmos in which tachyons could exist without disturbing the natural order observed in the physical realm. We can suggest a solution. Modern science uses measurements which unify linear extension and angular rotation in complex numbers. The complex plane, a complex dimension, has two rectangular components: A real axis, which corresponds to the ordinary numbers, and the axis rotated by 90° called the imaginary axis.

Measurements of spacetime use a metric with one real component, called temporal, and three imaginary axis, called spatial (at first, this metric was reversed until it was realized that the difference between plus and minus time was real, while plus and minus space was relative). There is little discussion about the fate of the missing components.

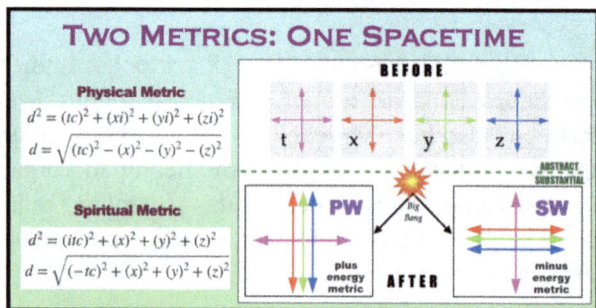

TWO METRICS: ONE SPACETIME

BEFORE

Physical Metric

$$d^2 = (tc)^2 + (xi)^2 + (yi)^2 + (zi)^2$$
$$d = \sqrt{(tc)^2 - (x)^2 - (y)^2 - (z)^2}$$

Spiritual Metric

$$d^2 = (itc)^2 + (x)^2 + (y)^2 + (z)^2$$
$$d = \sqrt{(-tc)^2 + (x)^2 + (y)^2 + (z)^2}$$

PW — plus energy metric
SW — minus energy metric
AFTER

We can imagine, however, that the Creator took four complex dimensions with its eight components. He assigned four—one real and three imaginary—as the metric of the physical world, and assigned the remaining three real and one imaginary components as the metric for the spirit world. As the square of an extension on the imaginary axis is a minus extension, the Pythagorean relations of separation in the two metrics are complimentary.

We have a complex spacetime in which the metric of the physical world is on one side, and the metric of the spirit world is on the other. (Note:

---

[1] https://www.forbes.com/sites/startswithabang/2019/02/12/why-supersymmetry-may-be-the-greatest-failed-prediction-in-particle-physics-history/#14bdb45869e6

there is no proof of this, it is just a suggestion.) If the negative energy tachyons are zipping along in the SW metric, they would not be expected to generate physical phenomena.

Postulating a four-dimensional complex spacetime with two complementary metrics provides a solution to one of the unsolved puzzles of quantum mechanics. In this well-tested theory, the empty vacuum has a tendency to transform into a pair production—say an electron-positron pair, a proton-antiproton pair, etc—and then transform back to empty vacuum.

This creation of positive mass is so brief that it does not amount to a Planck's Constant, the scientific measure of existence. Therefore, they are not real and are called virtual particles. For a brief moment—on the order of a billion, trillionth of a second—there is extra mass-energy in the vacuum. A volume of the vacuum—e.g. a

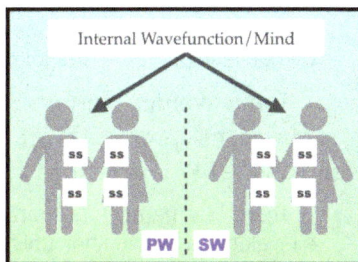
Internal Wavefunction / Mind

cubic millimeter—will have a small probability of having this extra energy in it. A cubic kilometer of vacuum— 10^{27} mm³—will on average contain 10,000 particle pairs of mass-energy.

A problem arises as there are a huge number of cubic kilometers in the universe, and all this virtual mass-energy flickering in and out adds up. Quantum theory suggests that the energy density of the universe should be on the order of 10^{100}; but it is observed to have a zero energy density. This is a huge disparity between theory and observation.

With the proposed two sides to spacetime, a solution would be that spacetime has a tendency to transform into a pair with positive energy as well as pair with negative energy. In this, the plus and minus neutralize and the calculated energy and observed energy match up.

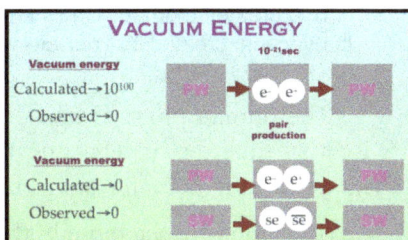
VACUUM ENERGY

The Spirit World

As discussed earlier, it is the internal wavefunction-mind that organizes interacting physical subsystems into a unified system with emergent properties derived from the Logos-natural law.

> Those who have had spiritual experiences testify that the incorporeal world appears as real as the world in which we live. The incorporeal and corporeal worlds together form the cosmos.[1]

The experience of both worlds is simple and familiar. An example is recorded in *Life in the World Unseen* as communicated to an earthly resident:

> We resumed our walk, and my friend said he would like to take me to visit a man who lived in a house which we were now approaching. We walked through some artistically laid out gardens, crossed a welt-turfed lawn, and came upon a man seated at the outskirts of a large orchard. As we drew near he rose to meet us. My friend and he greeted one another in the most cordial fashion, and I was introduced as a new arrival.[2]

The wavefunction-mind, being abstract, can equally organize the subsystems of the spirit realm into unified systems. From all accounts, the experience in the SW is similar to that in the PW.

> "If there is a parallel tachyonic universe, all velocities within that universe would be seen by us as greater than c, but would be seen as less than c by the inhabitants of that universe. Conversely, to the inhabitants of that universe, all velocities within our world would appear to be tachyonic."[3]

TWO REALMS: ONE COSMOS

PW | SW

Supporting this concept of a spirit realm composed of negative energy, we can look at the teachings of the many religions inspired by God, where the altruistic outward flow of energy is promoted for spiritual elevation, for spiritual life; while the inward flow of selfish gain is admonished as leading to spiritual decline, to spiritual death.

> Giving creates room for God's love to enter; the more room and the greater the vacuum created by your giving, the faster you will be filled by the flow of God's love.[4]

The human physical body on Earth is a result of genetics, while the human spiritual body, reflecting the personality, is generated by life in the PW, a product of human responsibility.

We conclude with the thought that modern science would be quite theoretically comfortable with a a non-physical spirit realm:

- If it is 72% of the Cosmos

---

[1] *Exposition of the Divine Principle*, p. 53

[2] http://www.ghostcircle.com/wp-content/uploads/2014/08/Anthony-Borgia-Life-In-The-World-Unseen.pdf, p. 17

[3] Oleksa-Myron Bilaniuk, Journal of Physics: Conference Series 196 (2009) 012021

[4] Sun Myung Moon, October 3, 1973, God's Will and World, God's Hope for Man

- If it has an antigravity effect
- If is composed of negative energy tachyons
- If it structured with supersymmetric entities
- If it inhabits the complementary metric
- If the wavefunction/mind organizes spirit-body subsystems.

Mind in SW and PW

We suggested earlier, that the spirit world had a set of fundamental particles just like the physical word has its quarks and electrons. This would be, in current theories, the supersymmetric set of particles that have been avidly searched for in vain. This makes sense in this view, they are all on the other side of spacetime. We will assume that these particles can be organized into structures in the spirit realm

Humans, as is very well established, are very sensitive to insubstantial qualities, such as love, beauty and truth. These are spiritual qualities that are invisible and insubstantial here in the physical wold, but are overriding and real in the spirit world.

We will now explore the concept of the mind, which is expressed in the physical structure, also being expressed simultaneously as a structure in spirit world. This would also embrace the wavefunction, which governs the structure of atoms, molecules and cells also have corresponding structures in the spirit realm.

The final assumption, is that the identical physical and spiritual structures—governed by the same wavefunction or mind—resonate together as single entity. Unsophisticated entities have unsophisticated properties, while sophisticated entities have sophisticated properties.

The advantage of this perspective is that it makes sense of otherwise unexplainable phenomenon.

A simple example is the Findhorn experiment, where plants on a Scottish garden responded to human love by developing abundantly and in unexpected ways. Plants have to have a simple spirit that responds to love, a quality that does not really make sense in a purely physical structure.

Moving up the levels of sophistication, we find that even reptiles—normally antithetical to humans—when raised in a loving environment can respond to love, if just barely. Visiting a friend in New Hampshire, I was horrified to see an enormous white python—in a large thick-glass chest—who was his pet. He offered to take the snake out of the chest so he could show

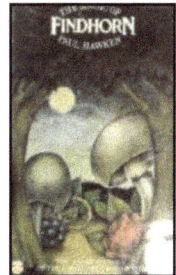

me how delightful a pet it was. As I had the distinct impression that the snake picked up my repugnance and would gladly squeeze me, I hastily, if politely declined and never returned to house again. Again we see that pythons are sensitive to love, and this is not a physical but a spiritual sensitivity.

Moving up the scale of sophistication, we see that spiritual aspect is considerably more sophisticated in the mammals, especially the pets we dote on. I am sure that I am alone in meeting dog, cat and horse lovers who are convinced that their pet has a real personality and is very responsive to love. Even mammals in the wild can be responsive to love, though you do have to be careful!

Also, be careful of telling a friend that their cat does not have a personality, or that their dog is not responsive to love: "They are just responding to you feeding them." That is the quickest way to get unfriended. Actually, food is love. The measure of your love for a companion is just how much you are willing to sacrifice for them: there is a wide range, from a moment to listen, through donating your salary, to the ultimate sacrifice of your life, the pinnacle of love.

Now pet owners sacrifice for their pets, their time, their money, their food. Some have even given their life attempting to save a pet.[1]

While humans seem to have exterminated all the hominids who gave rise to them, from our experience with their extant relatives—the Great Apes —I think we can assume that

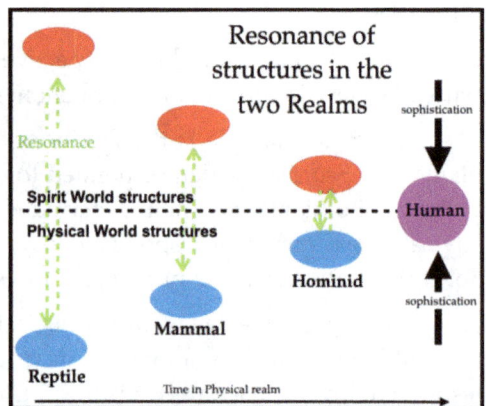

Resonance of structures in the two Realms

Resonance

Spirit World structures

Physical World structures

Reptile

Mammal

Hominid

Human

sophistication

sophistication

Time in Physical realm

---

[1] https://www.newsweek.com/dog-trainer-dies-house-fire-american-french-bulldogs-detorit-1683640

they were very receptive and responsive to love. They had a sophisticated spirit and brain.

All the cases discussed so far are mortal, the spirit structure dissipates as the physical brain dies along with the body. This is not so for humans. Both spirit and brain are at the acme of sophistication in someone like Jesus. This is when the dual entity has the emergent property of "I Am" which binds the spiritual structure together into an eternal personality. Even damaged humans—both spiritual and physical damage—have this eternal spirit.

In one of the few times that God directly communicated with spiritually broken humans, he used this precious moment to declare, just like us, God had that same quality

> And God said unto Moses: "I Am That I Am: and He said, Thus shalt thou say unto the children of Israel, I Am hath sent me unto you".[1]

Clearly, the "I Am" characteristic is of prime importance to our parental Creator!

We are now moving from considerations on a cosmic level, to dealing with the developments on Earth. While physically this is a tiny part of the

Years before Present	Event
4,500,000,000	Formation of Earth & Ocean
3,800,000,000	Origin of Life, LUCA
6,000,000	Last Common Ancestor of Humans & Chimpanzees
2,500,000	Origin of Hominids in Africa
100,000	Origin of Humans in Africa, Adam & Eve
70,000	Out of Africa diaspora
30,000	Extinction of Neanderthals
12,000	Agriculture, domestication plants and animals
4,250	First empire, the Akkadian Empire of Sargon
2,000	Jesus, Christianity, Rome
1,400	Islam
500	Newton, modern science
100	Electricity, medicine, WW 1 & 2, atomic energy, internet, UT

[1] Exodus 3:7–8, 13–14

universe, it is everything to us humans. This is an brief overview of the universe in terms of humanity:

6 • PRINCIPLED EVOLUTION

Many scientists in the physical sciences are coming to accept the concept of an Intelligent Creator, convinced by the fine tuning of the fundamental forces and constants that allow for life to exist at all, as discussed. The biological sciences, while dealing with the extravagance of life's burgeoning fecundity, have resisted this concept, and still cling to the contingent Darwinistic concept of random variation followed by the trials of natural selection.

In this book, we will compare Darwin's concept with the one presented by *Unification Thought* where the Intelligent Creator's Logos is progressively expressed over time.

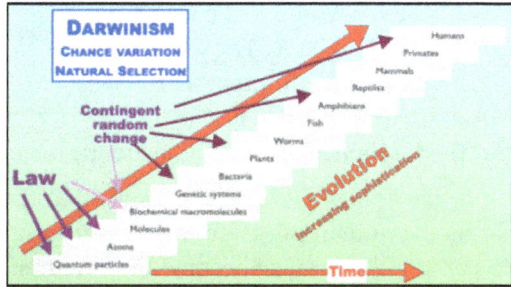

Darwinists accept that the basic realms of physics, chemistry and biochemistry are ruled by Natural Law, but the rest of the hierarchy is ruled by random accident. Neither the Origin of Life nor the subsequent evolution of Life's astounding complexity are ruled by law in Darwinism; life and its sensational evolution, in this view, are ruled by random contingency.

> The question of evolution's predictability was notably raised by the late paleontologist Stephen Jay Gould, who advocated the view that evolution is contingent and unrepeatable in his 1989 book *Wonderful Life*. "Replay the tape a million times ... and I doubt that anything like *Homo sapiens* would ever evolve again."[1]

A specific example is that when the monomers of proteins and nucleic acids are chemically synthesized in the laboratory they are racemic—they are equal amounts of right- and left-handed molecules.

The unexplained fact is that all Earth's living systems are not racemic —all proteins are assembled from left-handed amino-acids **L**, while all nucleic acids are assembled from right-handed nucleotides **R**. Gould's contingency suggests this was a contingent accident: life was not destined to be **L-R**, it could have been **R-L, L-L,** or **R-R**.

Unification Thought takes a different view. The Logos is hierarchical natural law that works on every level, from atoms to human. As discussed,

---

[1] https://www.sciencedaily.com/releases/2018/11/181108142323.htm

all the sophisticated entities that emerge over time have an abstract form inherited from the Logos. The Logos was generated by God before the Big-Bang, and in which all the creative work was completed. The physical world was to develop under the direction of the Logos, and only this, during the *indirect* dominion of God who was not in control.

Unification Thought suggests that if the tape of life's evolution was to be run again—such as on another planet—the result would be the same in fundamentals—proteins would all be **L** and DNA would all be **R**. The organisms could be as different as dinosaurs, lizards and birds are, but the fundamentals would be identical.

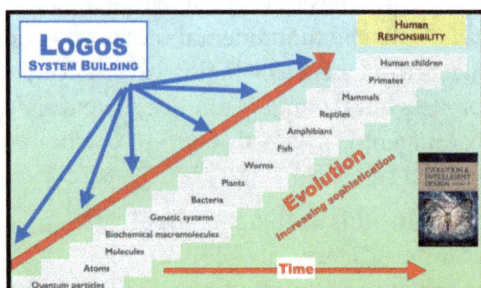

Unification Thought also states that the Logos only drives evolution as far as the human pre-teen: there is an extra portion of human responsibility to complete before we become the Children of God and enter the Direct Dominion of God's love.

The discipline known as *Intelligent Design* generates an excellent critique of the contingent accidental aspects of Darwinism and a strong case for a creative input along the course of evolution.

> Are life and the universe a mindless accident—the blind outworking of laws governing cosmic, chemical, and biological evolution? That's the official story many of us were taught somewhere along the way. But what does the science actually say? Drawing on recent discoveries in astronomy, cosmology, chemistry, biology, and paleontology, Evolution and Intelligent Design in a Nutshell shows how the latest scientific evidence suggests a very different story.[1]

Unfortunately, this discipline fails to suggest a way that this creative input arrives from God in the world in a scientifically acceptable manner. *Unification Thought* does not have this disadvantage and can supplement Intelligent Design's critique with a scientifically acceptable counterproposal.

Origin of Life

The Origin of Life has been disputed since Darwin suggested in a letter to J. D. Hooker in February 1871:

> "But if (and oh what a big if) we could conceive in some warm little pond with all sorts of ammonia and phosphoric salts, light, heat, electricity etcetera

---

[1] *Evolution and Intelligent Design in a Nutshell* (2020) Thomas Lo, Paul Chien, Eric Anderson, Robert Alston, Robert Waltzer; Discovery Institute, Seattle WA

present, that a protein compound was chemically formed, ready to undergo still more complex changes…"[1]

Many suggestions have arisen since to explain how life emerged:

Most are based on the assumption that cells are too complex to have formed all at once, so life must have started with just one component that survived and somehow created the others around it. When put into practice in the lab, however, these ideas don't produce anything particularly lifelike. It is, some researchers are starting to realize, like trying to build a car by making a chassis and hoping that wheels and an engine will spontaneously appear.[2]

In what can only be described as a revolutionary concept to emerge in the scientific press, the suggestion has emerged that:

[The] team found that the same starting chemicals can also make the precursors of amino-acids and lipids. All the cellular subsystem could have arisen through common chemistry, [they] concluded. The key is [called] 'Goldilocks chemistry': a mixture with enough variety for complete reactions to occur but not so much that it becomes a jumbled mess."[3]

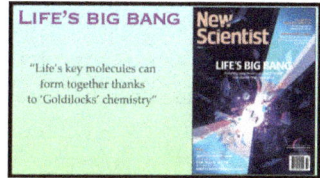

What is clear from the evidence, however, is that simple bacterial life was established in less that a million years after the molten Earth had cooled and an ocean established. While the origin of life is still a subject of intensive debate, its rapid emergence is in total agreement with the concept of the Logos acting on the internal wavefunction aspect of matter to make certain combinations highly probable.

Systematic Origins

As discussed, in modern science all physical entities have an intangible internal aspect as well as a tangible external aspect. For simple systems, such as atoms, the internal is called the wavefunction and the external is called the particle; for sophisticated systems, their names are mind and body.

The wavefunction determines the probability of coupling externally with subsystems in an interaction, while interaction changes the wavefunction. The Logos (natural law) works directly to determine the wavefunction, and how it changes in an interaction. When subsystems interact to form a higher system their wavefunctions merge becoming the new system's wavefunction, along with a set of emergent properties from the Logos expressed.

---

[1] https://www.darwinproject.ac.uk/letter/DCP-LETT-7471.xml

[2] Michael Marshall, Life's Big-Bang, New Scientist, August 8, 2020, p. 34.

[3] Michael Marshall, Life's Big-Bang, New Scientist, August 8, 2020, p. 37.

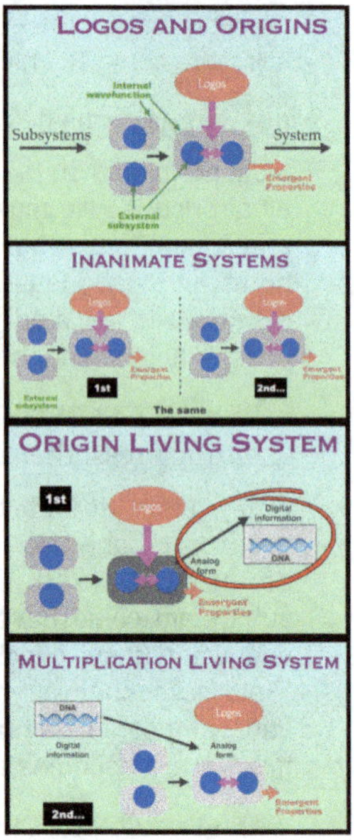

The Universe just after the Big-Bang was so hot that stable hydrogen atoms were impossible. Later, when the Universe had cooled by expansion, there were plenty of hydrogen atoms. So logically, there must have been the very first stable hydrogen atom in the Universe.

It is unheralded, however, as the zillions that followed originated in exactly the same way. The 1st, 2nd, 3rd etc atoms all appeared with natural law directing this multiplication of hydrogen atoms.

The origin of the first of a living system is qualitatively the same as for inanimate systems: the Logos is directly involved in the new analog form taken up by the interacting subsystems.

The great difference is that the information about this new analog form is recorded digitally as new information written to DNA.[1] The second of the system, and all all the others to follow, is not directed by natural law but directed by the stored digital information.

This 'writing to DNA' is anathema to the Fundamental Dogma of Darwinist Genetics where there is no writing of information, just reading of information that can only be altered by accident.

Digital information, Analog Form

There is a dual relationship in nature that is found only in living organisms, that of Digital Information and Analog Form. In order to begin to understand this duality in living systems, we will look at this well-established duality in computer science. First we will consider the great similarities between the two. Then we will look at the major difference between them.

In Darwin's day, there was no understanding of how a simple seed or egg could develop over time and transform into a mature organism such as a tree or a chicken. So there was no understanding of why offspring reflected their parents, or why there was variation between siblings.

It is the stored digital information, not the direct action of the Logos, that brings the subsystems together in the 2nd, 3rd, etc., generations. The

[1] It is thought that in the beginning of life, it was RNA that was the information store. This is a minor point as DNA is just waterproofed RNA, a few dabs of hydrophobic oil added to one of the bases, and the backbone stripped of hydrophilic hydroxyl groups.

resultant analog form resembles the form in the Logos, so the same emergent properties are present.

We find a qualitative difference between non-living systems—where the origin of every system directly involves the Logos—and living systems where the Logos is directly involved in the origin of the very first, but not in the subsequent generations where digital information plays a direct role.

Condensing a long history of exploration, scientists found that all the analog qualities—such as blue eyes— that were passed down a lineage (technically: the phenotype) were encoded as digital information encoded in the DNA content of chromosomes (the genotype).

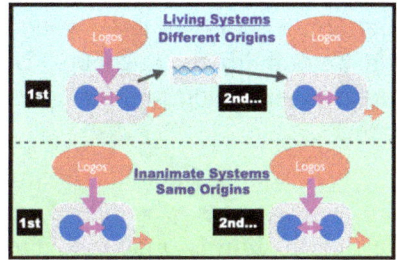

There are also analog qualities passed down a lineage that, while managed by DNA information, cannot be created de novo. Examples are the bi-lipid membrane, the ribosome and the centrosome—no extant organisms can make a ribosome without a ribosome, a bi-lipid membrane without a bi-lipid membranes, or a centrosome without a centrosome. But it is digital information that must be decoded into analog form that is, by far, the most important aspect of inheritance.

The relation between digital information and analog form was unknown in Darwin's day, Now, however, we are in the Age of Computers, where the connection between these two is familiar.

We will start by noting the similarities between digital information manipulation in living and computer systems and conclude with the major difference between the two in current thinking.

Living and Computer systems

The flow, manipulation and decoding of digital information in the cell is similar to that in a computer. So we can use the insights of computer science to help us understand the vastly more sophisticated living systems. In computer systems, the digital information is stored in a binary code of complementary bits **0 & 1.** The basic manipulation is **NOT**, where **NOT 0 =1** and **NOT 1 = 0.**

The manipulation in computers involves sets of 8 bits, called a byte. The external form of these are various—magnetic poles, pits in aluminum, holes in paper type, radio waves, sound waves, etc.,—but the digital information remains the same.

In living systems, the digital information is stored in a di-binary code, the two pairs of complementary dibits **00** & **11** & **01** & **10** whose most basic manipulation is **NOT,** where **NOT 00 = 11** and **NOT 10 = 01** and vice versa. In genetics, the **NOT** form of a nucleic acid is called its complement. The manipulation involves dibits in sets of three, called triplets. The external form of these can be the chemical bases, guanine-cytosine (**G-C**) and, adenine-thymine/uracil (**A-T/U**) in DNA, mRNA, tRNA, etc., but the digital information is the same.

In computer systems, digital information is organized into Apps that perform a variety of tasks when called into action. The app is processed on the Central Processing Unit (CPU) which manipulates a variety of inputs and outputs an analog form, such as this article, a Netflix movie or a Beatle's song.

A similar situation occurs in living systems, where the digital information is organized into Genes that perform a variety of tasks when called into action. The gene is processed by the RNA-protein matrix of the nucleus (the cell's CPU) which manipulates a variety of inputs and sends an output to a ribosome which decodes it into a protein with a particular analog activity.

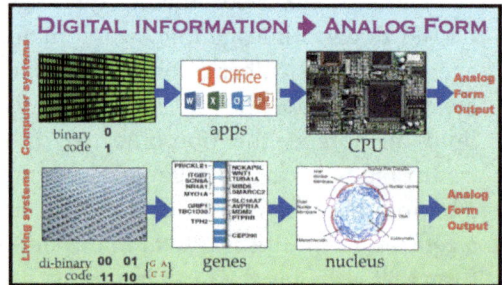

A simple example is the SDY gene on the male Y chromosome. A few weeks into fetal development, the gene is activated for about and hour, generating a protein that starts processes that cause a fetus to develop male gonads (testes) and prevent the development of female reproductive structures (uterus and fallopian tubes). This gene is never called upon again in the life of the male.

In early computer systems, such as the Apple II, the digital information was processed one byte at a time. The digital information was almost all in the ASCII code which assigned a particular byte to a particular alphanumeric character. Examples being: **A = 01000001, B=01000010.**

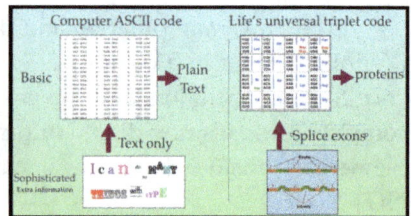

The output was a stream of text sent to a printer having just one font, usually `Courier`: **The analog output of early computers looked like this.**

In the simple, earliest forms of life, such as the bacteria, the digital information is also almost all in the Triplet Code which assigns a particular triplet to an amino-acid. Examples: **Glycine = GGU**, **Arginine = AGG.** The information is passed directly to a ribosome for translation into an array of amino-acids, which folds into a protein with an analog task to perform.

An example is the protein enzyme, catalase, a common enzyme found in nearly all living organisms exposed to oxygen (such as bacteria, plants, and animals). It catalyzes the decomposition of hydrogen peroxide to water and oxygen. One catalase molecule can convert millions of hydrogen peroxide molecules to water and oxygen each second.

The modern M2 MacBook Air I am writing on is a 64-bit computer, manipulating 8 bytes at a time. In this document the ASCII code is still in use by one of the eight bytes, but the other seven contain extra information in a different code allowing typographical idiocy such as:

I c a n do so mANy THINGs with tYPE.

I can strip away all that extra information, however, by choosing *Text Only* and the output is, "I can do so many things with type."

In sophisticated living systems, eukaryotes such as humans and spinach, genes still contain islands of Triplet Code information, called Exons, surround by digital information in a different code, called Introns. Just like the Text Only command, the intron RNAs are excised and the exons RNAs spliced together for export for translation to protein by a ribosome.

The intron RNA snippets join the dozens of different kinds of RNA at work in the nuclear matrix contributing, in an unknown way, to the functioning of the cell's CPU. The RNA in the nucleus certainly seems to play a central role in the structure and functioning of the nucleus. For example:

> Recent evidence… that long non-coding RNAs in particular may play a central role as a kind of 'scaffolding' that ties different regions of the genome together both structurally and also in terms of function. That such RNAs, by virtue of their sequence but also 3D shape, can bind DNA, RNA, and proteins, makes them ideal candidates for such a role. Importantly, the scaffolds that they form seem highly dynamic, and this may be a key factor in the regulation of gene activity in a global fashion across the genome.[1]

In the early days of genetics when only the simple and direct methods used in bacteria were understood, all the non-coding DNA that was not translated into protein was labeled as "Junk DNA" and genes that had no

---

[1] Parrington, John. *The Deeper Genome: Why there is more to the human genome than meets the eye* (pp. 145-146). OUP Oxford.

known function were labelled as "selfish" as their only purpose seemed to be propagating themselves. Nowadays, the situation has radically changed. It is now understood that dozens of types of non-coding RNAs are transcribed from DNA to run the workings of the nucleus, the CPU for digital information.

A recent list had over 70 types of RNA.[1] The RNAs can work directly with simple enzymatic activities (ribozymes) such as splicing exons. Or the RNA complex can order up proteins which are transported back to the nucleus with direct or assistant activities to the RNA complex activities.

Only 5% of the digital information on DNA codes for proteins, the rest is command and control functions that are now coming to light. This is similar to a sophisticated MS Word document where only 3% of the digital information is ASCII code.

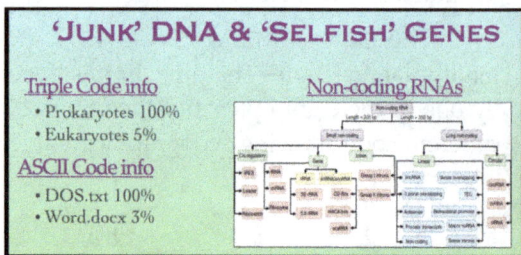

'JUNK' DNA & 'SELFISH' GENES

Triple Code info
• Prokaryotes 100%
• Eukaryotes 5%

ASCII Code info
• DOS.txt 100%
• Word.docx 3%

Non-coding RNAs

Reading and Writing

So far, we have illustrated the similarities between living and computer systems where the management of digital information is concerned. Now to discuss the one great difference between computer science and current evolutionary theory.

The combination of Darwinism and Genetics is called the Modern Synthesis. Its fundamental dogma is that information flow is one-way: DNA to RNA to Protein. This is where the Modern Synthesis departs from computer science.

We take it for granted, that the Hard Drive, where the computer stores its digital information, allows for us to **Read** information from the disk, as well as **Write** information. Every time we buy a new app, it is written to the hard drive until we need it.

The fundamental dogma of the Modern Synthesis is that DNA—the hard drive storage for the cell—is **Read-Only**, from digital DNA to analog form. In this view, there is no **Write** function. The only way the digital information stored in

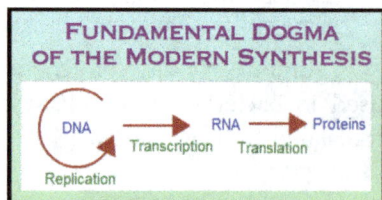

FUNDAMENTAL DOGMA
OF THE MODERN SYNTHESIS

DNA ⟶ RNA ⟶ Proteins
Transcription   Translation
Replication

---

[1] https://en.wikipedia.org/wiki/List_of_RNAs

DNA can alter is by random chance-and-accident mutation along with errors in duplication. This is the central dogma of genetics: Information flows from DNA to RNA to Proteins to Analog Form in the cell. DNA digital Information is accumulated over time by random alterations, and the accidental analog forms it generates are selected for usefulness in survival and reproduction.

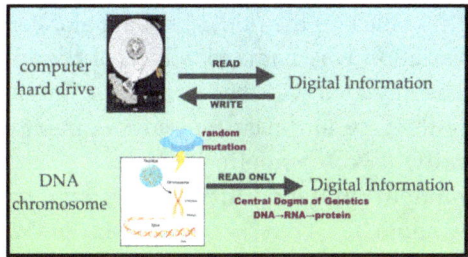

This is as weird as if Bill Gates explained how they created Microsoft Windows: "We subjected hard drives to X-rays and hammer blows. The random bits generated were then tested for usefulness in running a PC." (Some Mac enthusiasts still believe this rumor.)

As any computer programmer will assert: dogma or not, this is no way to write useful digital information. This Read-only dogma remained unchallenged until work with RNA viruses uncovered enzymes that could copy RNA onto DNA, the enzyme Reverse Transcriptase.

Most unexpectedly, the human genome was found to have ~600 reverse-transcriptase-like genes. These are currently considered remnants of RNA-viral infections in the pre-human lineage and, along with introns and other non-coding DNAs consigned to Selfish status.

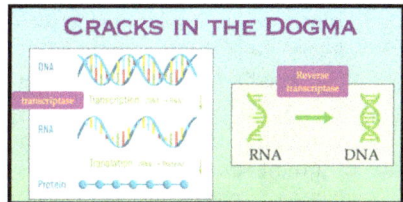

Just because no one has see these genes in action, however, does not necessarily imply that they are never run. Some genes, like an esoteric app, are only needed occasionally, as illustrated by the *Sex Determining Gene* on the Y-chromosome which is only active in a fetus for an hour, and never again in the lifetime, yet its momentary transcription sends the fetus down the development path to male rather than the default female.

This was the first crack in the Read-only dogma. The next was the currently emerging science of Epigenetics stimulating the contemporary revival of Lamarck's ideas.

> This striking personage in the history of biological science, who has made such an ineffaceable impression on the philosophy of biology, certainly demands more than a brief [tribute] to keep alive his memory.[1]

---

[1] Packard, A. S. (1839-1905) *Lamarck, the Founder of Evolution His Life and Work*. Public Domain.

Back in the mid-1800s, a major competitor to Darwin's view of random variation was Lamarck's idea of the inheritance of acquired characteristics. The key idea was that learnt wisdom about how to deal with the challenges of thriving in the current environment could be passed onto offspring. Basically, two competing ideas: random mutation or learnt wisdom.

Lamarckism fell out of favor due, to my mind, a stupid experiment of cutting the tails off of mice for generations, and noting that they learnt nothing to pass on down their lineage.

Lamarck's basic idea reemerged in 2013 when researchers recorded effects on the grandchildren of survivors of a dreadful famine, the Nazi 'Dutch Hunger Winter' in which more than 20,000 people died of starvation.

Because of the excellent health-care infrastructure and record-keeping in the Netherlands, epidemiologists have been able to follow the long-term effects of the famine. Their findings were completely unexpected.... some of these effects seem to be present in ... the grandchildren of the women who were malnourished during the first three months of their pregnancy. So something that happened in one pregnant population affected their children's children. That raised the really puzzling question of how those effects were passed on to subsequent generations.[1]

Such findings initiated the new science of Epigenetics. This new science, still in its infancy, is researching the ways that gene expression can be altered in ways that can pass down the generations.

Although the discipline of epigenetics is only decades old, it is already intimating that the random chance and accident mechanism driving evolutionary change in Darwinism will one day be replaced by the learning mechanisms postulated by Lamarck. Such insights are [epigenetically] passed down a lineage as ancestral wisdom about success in the created world, as suggested in a recent book on the potential impact of this new aspect of evolution.[2]

---

[1] http://www.naturalhistorymag.com/features/142195/beyond-dna-epigenetics

[2] Peter Ward, *Lamarck's Revenge: How Epigenetics is Revolutionizing Our Understanding of Evolution's Past and Present*, Bloomsbury Publishing, NY, 2018

Darwin was not at all a fan of Lamarck's idea of going beyond natural selection to a more constructive view. It is recorded that he said:

> "Heaven forfend me from Lamarck's nonsense of a 'tendency to progression,' 'adaptations from the slow willing of animals,' etc. But the conclusions I am led to are not widely different from his; though the means of change are wholly so."[1]

Actually, if we replace Lamarck's somewhat misty intuition of a 'tendency to progression' with the perspective derived earlier from Quantum Physics: That the external structure of a system is a reflection of the internal wavefunction.

The structure and development of the internal wavefunction is a ruled by Natural Law combined with the external interactions (coupling with its subsystems) of the system. The internal wavefunction governs the probability of what the external system will do and interact

So natural law indirectly influences the mind of systems to move towards the goal of Natural Law, the ascent of the hierarchy listed in an earlier chapter. This is a modern explanation of exactly what Lamarck's 'tendency to progression' actually is, and how it influences probability.

Well before epigenetics as a field of study and the writing to DNA, experimental support for Lamarckism emerged. In September 1988, a paper on bacterial genetics was published in a prestigious journal[2]. This paper appeared to contradict the fundamental tenet of neo-Darwinism evolutionary theory: the principle that mutations, that the source of genetic variation, occur randomly and that the direction of evolution is supplied by natural selection—the "survival of the fittest.

This paper promoted a great debate:

> The findings appeared to support the discredited Lamarckian theory of evolution—the starved bacteria weren't growing long necks but, just like Lamarck's imaginary antelope, they appeared to be responding to an environmental challenge by generating heritable modifications: mutations....

> The finding also appeared to contradict what is sometimes called the central dogma of molecular biology: the principle that information flows only one way during transcription, from DNA out to proteins to the environment of a cell or organism. If [these] results were right, then cells must also be capable of reversing the flow of genetic information, allowing the environment to influence what is written in DNA.[3]

---

[1] quoted from (Darwin's Life and Letters, ii., p. 23, 1844.) in: Packard, A. S (1839-1905). *Lamarck, the Founder of Evolution His Life and Work.* Public Domain.

[2] J. Cairns, J. Overbaugh & S. Millar, "The origin of mutants," Nature, vol. 335 (1988), pp. 142–5.

[3] McFadden, Johnjoe; Al-Khalili, Jim. Life on the Edge: *The Coming of Age of Quantum Biology* (pp. 221-222). Crown/Archetype.

This directed change is inconsistent with theory, a point noted by Darwinism's most vocal current advocate, Richard Dawkins:

> Evolution is very possibly not, in actual fact, always gradual. But it must be gradual when it is being used to explain the coming into existence of complicated, apparently designed objects, like eyes. For if it is not gradual in these cases, it ceases to have any explanatory power at all. Without gradualness in these cases, we are back to miracle, which is simply a synonym for the total absence of explanation.[1]

Speciation mechanism?

The chromosomes we see in pictures of cell division are where the DNA (the A is acid) is tightly wound onto spools of alkaline histone proteins, called nucleosomes. These, in turn, are coiled, and supercoiled into the dense form of a chromosome. This is like wrapping ten miles of cotton thread onto tiny spools, coiling these into string, and these into ropes, and these into a foot-long braid!

The nuclear mechanisms can only access the DNA to transcribe its information onto RNA when it is liberated from this condensation. Control of this condensation is crucial. Access can be stimulated by adding acetyl groups to the histones, among other methods; access can be inhibited by adding methyl groups to the cytosine nucleotides, among other methods.

Epigenetics is currently detailing the writing, the reading and erasure of epigenetic information recording the current state of the cell and how it is doing. Just a few of the many mechanisms being explored currently are illustrated here.

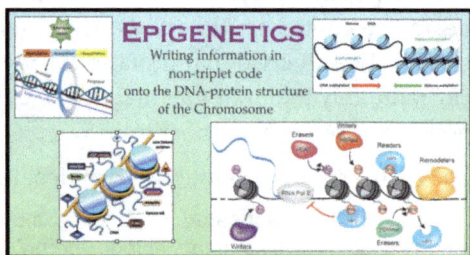

EPIGENETICS
Writing information in non-triplet code onto the DNA-protein structure of the Chromosome

Returning to our computer science analogy, the information epigenetically encoded—and the code has yet to be fully understood—can be likened to notes added to a printout of a program under development. Comments such as "this is not working well" or "the choices here can be simplified" cause the writer to adjust the program before it is sent to the compiler.

The Compiler converts the high-level language into the binary bits of machine code that the app that will run on the CPU. In living systems, these epigenetic notes are accumulated down a lineage and help each individual prosper. As the information mounts up, eventually it is sufficient to cause a speciation event where the short term memory (epigenetic) is converted into

---

[1] Dawkins, R. (1995) *River Out of Eden*, Basic Books, New York, p. 83.

long-term memory. The equivalent of the compiler adjusts the machine code, the genetic digital information stored in the DNA.

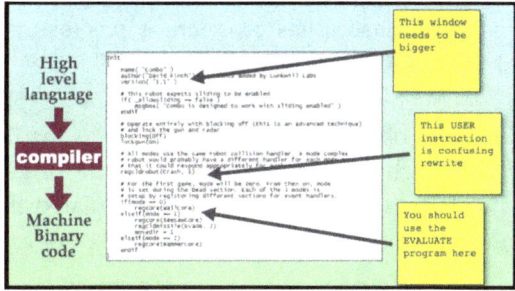

Life's 'compiler' is the nuclear machinery that runs the creation of the sexual gametes and the recombination of chromosomes in Meiosis. A normal (somatic) cell has two copies (diploid) of each chromosome, one from dad, one from mum. In regular mitosis, the chromosomes are doubled, then segregated into two new diploid somatic cells.

Meiosis starts similarly with duplication of the chromosomes in a diploid germ cell. The four copies of tetraploid DNA are now mixed and matched by the nuclear machinery in what is called recombinant crossing over. The four chromosomes are now segregated into four single copy (haploid) sex cells (gametes) with chromosomes where dad and mum's contributions have been mixed.

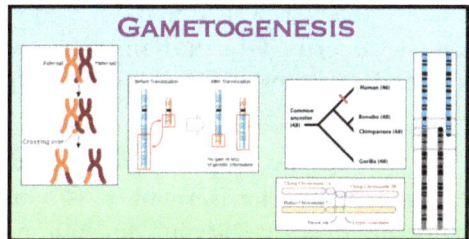

It is quite possible that the reverse transcriptase ability is called upon to copy RNA onto DNA at this time of transformation. What is now beginning to be explored is that epigenetic information influences these manipulations of tetraploid DNA:

> The assembly of [tetraploid DNA] is driven by the combinatorial action of many factors including histones, their modifications, and [epigenetic] DNA methylation.[1]

It is well established that a major manipulation of chromosomes occurred in the lineage leading to humans as the human #2 chromosome is clearly generated by fusing two great ape chromosomes together.[2]

Somewhere in the lineage that led to humans the two chromosomes were fused together. The complex cellular mechanisms organizing and regulating the chromosome transformations in meiosis are currently being explored.

---

[1] https://www.ncbi.nlm.nih.gov/pmc/articles/PMC4830869/

[2] https://genetics.thetech.org/ask/ask229

Our understanding of the molecular mechanisms governing meiotic recombination has considerably progressed these last decades, benefiting from complementary approaches led on various model species.[1]

Such epigenetic transformations are probably at work in the microevolution as observed by Darwin, such as the finches on the Galapagos Islands. He thought this microevolution—the origin of variety—was random, not directed, variation. When he extrapolated this to macroevolution, the origin of species, he also included the concept of random variation.

The sudden changes in the stored genetic information that mark speciation, informed by epigenetic information, explains a fact that has bedeviled Darwinistic thinking. Darwin's theory states that the gradual accumulation of variation will eventually be so extreme that two species, not one, come to exist. Darwinism predicts there should be gradual changes between extant species, as well as in the fossil record. Gradualism should be the norm if Darwin is correct.

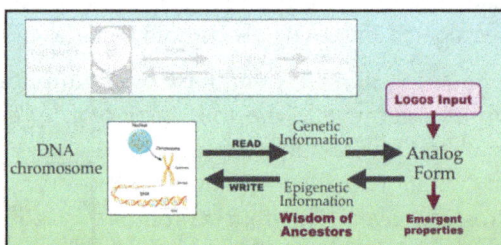

The science adage—many a great theory has been undone by facts—is exemplified here for Darwinism. Gradualism is prevalent in microevolution, as exemplified by Darwin's finches on the Galapagos Islands. Gradualism, however, is absent in both in extant species and the fossil record. There are, for instance, no Darwin's sparrows or blackbirds on the Galapagos, just finches.

This expectation was not realized; fossil species were found to be all distinct and appearing suddenly in the record. It took a great deal of nimble thinking to incorporate this fact into Darwin's theory.

In the 1970s, evolutionary scientists Gould and Eldridge proposed an explanation, which they called "punctuated equilibrium." That is, species are generally stable, changing little for millions of years. This leisurely pace is "punctuated" by a rapid burst of change that results in a new species and that leaves few fossils behind.[2]

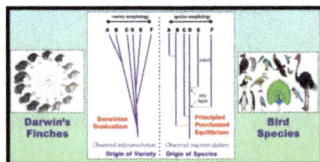

While is was a good effort, they were accused of being 'anti-Darwinian.' Rather than relying on the rapid disappearance of intermediate forms,

[1] https://pubmed.ncbi.nlm.nih.gov/27180110/

[2] https://en.wikipedia.org/wiki/Punctuated_equilibrium

Principled Evolution suggests that the punctuation events are a result of a new input from the Logos without any intermediate forms.

We see that genetics is uncovering what our experience with computer science insists must exist: a **Write** as well as a **Read** process for stored digital information.

The natural environment, which is a reflection of the lawful Logos, has an input to the analog form. This is conveyed by epigenetic information written onto the DNA. This, in turn, controls the development of the genetic information.

We conclude with this thought: The Modern Synthesis, the combination of genetics and Darwinism, is a paradigm ripe for replacement by a Postmodern Synthesis of epigenetics and Lamarckism. What might be called Principled Evolution, replacing the old idea of evolution by random-chance-and-accident mutation and mistakes in copying.

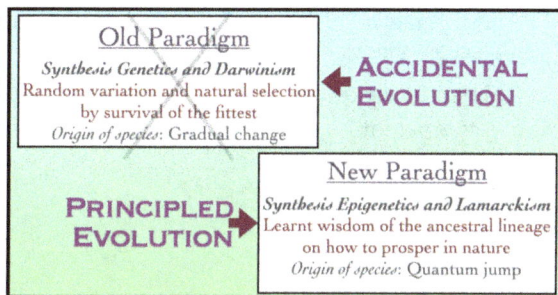

Old Paradigm	
Synthesis Genetics and Darwinism Random variation and natural selection by survival of the fittest *Origin of species*: Gradual change	◀ **ACCIDENTAL EVOLUTION**
New Paradigm	
PRINCIPLED EVOLUTION ▶	*Synthesis Epigenetics and Lamarckism* Learnt wisdom of the ancestral lineage on how to prosper in nature *Origin of species*: Quantum jump

As is often the case, it might be that the true situation is a blend of both perspectives on evolution:

Eugene Koonin has recently argued that evolution may follow a 'two-phase process, with the first phase being the Lamarckian epigenetics and the second phase Darwinian selection of mutations.' His proposal is that this would be akin to 'probing the waters…with epigenetic adaptation followed by the long-term genetic inheritance of the same adaptation should the challenge prove to be long-lasting'. If true, Koonin believes that this 'defies the common belief that evolution has no forecast'.[1]

For epigenetics to have a role in evolutionary speciation, any intelligence garnered by the somatic body would have to be sent, and processed, by the germ cells of the next generation. Suggestions have been made:

How could stress lead to changes in microRNA levels in the sperm? One possibility is that stress hormones circulating in the blood make their way to the testicles, and trigger expression of microRNAs via stimulation of surface receptors on the sperm. However, an even more direct potential route has been recently identified, since microRNAs contained within [circulating]'exosomes' … It is possible, therefore, that microRNAs produced elsewhere in the body, for instance, the brain, could subsequently end up in

---

[1] Parrington, John. *The Deeper Genome: Why there is more to the human genome than meets the eye* (p. 183). OUP Oxford.

the sperm and in the fertilized egg and embryo, providing a direct connection between the brains of one generation and the characteristics of future ones.[1]

In 1871 one of Darwin's critics, St. George Mivart, listed his objections to Darwin's theory, many of which are surprisingly similar to those raised by modern critics:

> What is to be brought forward (against Darwinism) may be summed up as follows: That "Natural Selection" is incompetent to account for the incipient stages of useful structures. That it does not harmonize with the co-existence of closely similar structures of diverse origin. That there are grounds for thinking that specific differences may be developed suddenly instead of gradually. That the opinion that species have definite though very different limits to their variability is still tenable. That certain fossil transitional forms are absent, which might have been expected to be present…. That there are many remarkable phenomena in organic forms upon which "Natural Selection" throws no light whatever.[2]

The ideas explored here replace incompetent random variation with a highly-competent Logos of natural laws.

---

[1] Parrington, John. *The Deeper Genome: Why there is more to the human genome than meets the eye* (p. 183). OUP Oxford.

[2] Mivart, St. G. (1871) *On the Genesis of Species*, Macmillan and Co., London, p. 21.

7. DIGITAL AND ANALOG

In Unification Thought (UT), one of the fundamental aspects of God is that of dual characteristics. Eastern philosophy recognizes the external dualities of yang and yin, male and female, positive and negative, etc. UT asserts that there is the even more fundamental duality of internal character—sung sang—and external form—hyung sang; such as mind and body at the most sophisticated levels, wavefunction and particle at the most basic levels.

Similar to this internal/external duality is the relation between that of abstract information and its substantial expression as analog form. Information is abstract, it is invisible and intangible. It can exist in the mind but, to have any influence it has to be expressed in an external form.

Two Biblical examples are the Ten Commandments revealed to Moses that were inscribed on the tablets of stone; and Jesus as the Word made Flesh. Adherents of three major religions—Judaism, Christianity and Islam—are called the People of the Book as they depend on written information to guide them.

Newspapers, scientific reports, billboards, etc, add images and charts, etc, to enhance words to improve the communication of information, be it a cure for Covid or pill to lose weight. The mode of expression can be on paper, on a TV set or a computer screen.

All these expression are technically all analog, they involve simple shapes—the alphabet—or complex forms—pictures of people in nature. Analog forms are the stuff of everyday life.

It was only within the last century that a new and very different way of expressing information was developed that did not involve analog form but digital information expressed as analog form.

Digital Information

As is often the case, God had long ago used this duality in creation before it was discovered by us humans. This aspect of creation is not universal, however, it only plays a role in living entities. It plays no role in non-living entities such as minerals and stars. This aspect of life is what we call digital information.

This aspect of life was uncovered in the last half of the 20th century and its significance roughly paralleled the technology of the computer. For both living systems and computer systems rely on the utility and density of digital information for their innermost workings.

The results of this innermost level are expressed in the outermost levels as analog form—cells in living systems, computer screens in computer systems. Density, in this realm, means compactness. For example, the Library of Congress holds about two hundred million items—books, reports, etc.—in its repository. This analog store contains about 15 petabytes (millions of gigabytes) of digital information. While it would take years and great expense to transfer all 200 million books from Washington to San Francisco, it would take only about 25 minutes and little cost to transfer all the digital information over current commercial optic cables—and fractions of a second in laboratory conditions!

> The fastest fiber-optic cable speed offered in a business gigabit network service today is 10 Gbps. The fastest fiber-optic cable speed ever recorded is 1 petabit [million Gb] per second, recorded by the Japanese company NTT in 2012.[1]

Unlike earlier ages that had little experience with digital information, we have now lived with computers for decades, and the use of digital information is increasingly common. This is fortuitous as we can apply many of concepts developed in the digital world of computers to the digital processes being uncovered in living systems.

One of the most significant concepts used in computers is that of mathematical abstraction. The digital information used by computer technology can take many substantial analog forms—holes in paper, pits in metal, patterns of ink, magnetic poles, electric voltage and frequency, radio waves, sound frequency, to name but a selection. Computer science and theory ignores these externals and focuses on the mathematical abstraction of the binary pair of ones and zeros written in all these different forms—such as hole or no-hole, N or S magnetic pole, on or off, open or closed, etc.—encoding the binary digits (bits) of **1** and **0**.

The most basic abstraction is the relation of 'ones complement' between this pairs of bits. This is the logic operator, **NOT**, where 1=**NOT** 0, 0 =**NOT** 1. It is irrelevant to theory how this logical manipulation is accomplished, only the logical manipulation is important. This is manipulation of a single input to a single output.

Other logical manipulations have two inputs and a single output, an example being the 'exclusive or' logical **XOR** where the output is a 1 if the

---

1 https://enterprise.spectrum.com/support/faq/internet/what-is-the-speed-of-fiber-optic-cable.html

inputs are different, 0 if they are the same. There are many logic operators such as this, some taking 3 or more inputs such as **AND, NAND,** etc.

In the earliest days of computers, these manipulations were performed by vacuum tubes, nowadays they are performed by silicon logic gates, each with its own symbol in circuit diagrams. It might be surprising to know that all the work of computers is accomplished by concatenating circuits of such logic gates, on a silicon chip.

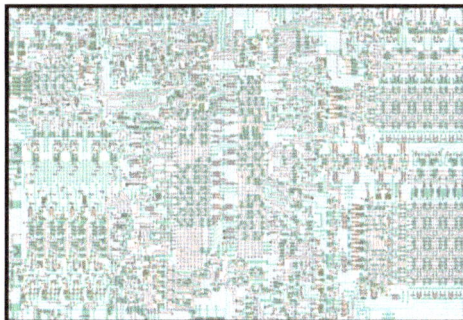

Genetics, the study of digital information in living systems, is still in its infancy and has not yet moved from substantial chemicals to abstracted logic. The set of four symbols for the four chemical nucleotides on which information is manipulated by RNA are [G-C, A-U] and when stored on DNA they are [G-C, A-T] (T being U 'oiled' by a methyl group attached). One thing that Genetics has not learnt from computer science is to ignore the analog expression and focus on the digital information.

The triply bonded analog C-G pair is as the digital pair 00-11, while the doubly bonded analog A-T/U pair is the digital 01-10 pair. Rather than the single bits of the computer we have the double dibits of genetics.

Their relation is exactly that of the **NOT** logic operator:

G =**NOT** C [00 = **NOT** 11],   C = **NOT** G [11= **NOT** 00]

A =**NOT** U/T [10 = **NOT** 01], U/T = **NOT** A [01= **NOT** 10]

While a computer needs a silicon logic gate to NOT, the nucleic acids do it automatically as the base-pairs cling together with hydrogen bonds. DNA is composed of a double helix of complementary strands bound together by the H-bonds, one strand of digital information is the template strand, the other is the coding strand. The relation between the digital information on the two strands is simply that of *template* = **NOT** *coding*, and vice versa..

A similar abstraction is when the two strands of the DNA helix are separated, each is copied as a **NOT** strand, and the two sets of strands recombine as two identical helixes. A set of proteins now perform an **XOR** on each step of the helix, if the output is not a 1—a proper pair—an enzyme corrects the situation.

ARRAYS OF ARRAYS

One thing that computers and genetics have in common is the manipulation of digital bits and bibits in basic arrays that are then linked together in larger arrays.

| A 01000001 | I 01001001 |
| B 01000010 | J 01001010 |

ASCII computer code
Triplet genetic code

GGA	CCT	GAG	CTC	GAA	CTT
AGA	TCT	AAG	TTC	AAA	TTT
GGC	CCG	GAT	CTA	GAC	CTG

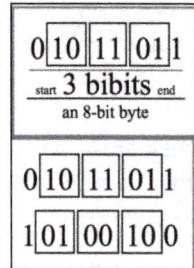

For computers, the basic array is a byte of eight bits. In genetics, the basic array is a triplet of three bibits.

One thing that is different is that while bytes are always read in one direction, from left to right, in genetics the arrays are polarized. The carbons in the ribose molecule that links the nucleotides are numbered 1' - 5', with a phosphate on the 5' and an hydroxyl on the 3' positions. Reading, writing and assembly of nucleic acids always runs from the 5' start to the 3' finish, and when strands bind they are always running in complementary directs: the 3' end of one strand is matched with the 5' start of the other.

| 0 | 10 | 11 | 01 | 1 |

start **3 bibits** end
an 8-bit byte

A conveniently way of indicating a polarized triplet dibit of genetic information in a computer 8-bit byte is to indicate the 5' start with an initial zero and the 3' end with a one. Then, when complemented, the correct orientation is automatically preserved.

| 0 | 10 | 11 | 01 | 1 |
| 1 | 01 | 00 | 10 | 0 |

Translation

The simplest form of life, the bacteria, and the 8-bit early computers had much in common when it comes to translating digital code into analog form.

Even the simplest of word processing computer—my first was an 8-bit Radio Shack Model 100 with 25k! of memory—was superior to any high-end typewriter in its liberation from White-Out and its capacity for endless rewrites.

8 00111000	G 01000111
9 00111001	H 01001000
A 01000001	I 01001001
B 01000010	J 01001010
C 01000011	K 01001011
D 01000100	L 01001100
E 01000101	M 01001101
F 01000110	N 01001110

The Model 100 could store 25 thousand letters of the alphabet using the ASCII code where each byte stood for one letter. The illustration is a section of this table.

There are 256 possible bytes, more than sufficient for the upper and lower alphabet 10 numerals, space carriage return, () , . : " ? ! @ # , etc., along with a few simple commands such as {new paragraph}. All the content of memory was taken up by this ASCII code. Using this code, the ASCII information could be output by a translation mechanism as paragraphs of text.

One significant innovation connected the computer to a Selectric and ran the typewriter! Astounding in its time! But limited to an output with:

a single font, a single size, a single monospaced font such as 10 point Courier.

In the next section, we describe how proteins are responsible for almost all of the substantial analog form of living systems. The capacity of protein to generate and manipulate analog form, however, is utterly dependent on the linear sequence of amino-acids in their primary structure.

In a similar way to the Model 100, the genetics of bacteria translate a linear sequence of triplets into a linear sequence of amino-acids, the primary structure. There are 64 possible triplets while only 20 amino-acids, so the triplet code is degenerate in that there can be many codes for a single amino-acid as well as a few simple commands such as [Start] and [Stop]. The table shows a part of the Triplet Code for some of the 20 amino-acids. This code, with a few minor tweaks, is universal and is used by all living organisms from bacteria to oak trees to humans.

CAU		CGU	
CAC	His	CGC	
CAA		CGA	Arg
CAG	Gln	CGG	
AAU		AGU	
AAC	Asn	AGC	Ser
AAA		AGA	Stop
AAG	Lys	AGG	Stop

In all living things the synthesis of proteins with a specific primary structure is basically identical. In genetics, when digital information for protein assembly is read from the _coding_ strand of DNA it is transcribed as its **NOT** complement onto a strand of mRNA, with the sequence identical to the _template_ strand.

The digital information on the mRNA is complemented onto an array of transfer RNAs (now with the _coding_ sequence) which read the information a single triplet at a time —the Triplet Code—and each tRNA has a specific amino-acid attached to it. This is the final step facilitated by ribosomal RNA.

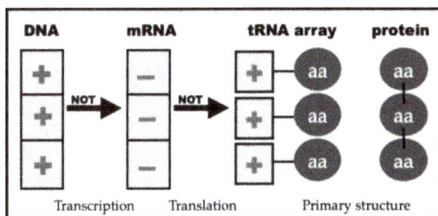

DNA — mRNA — tRNA array — protein

Transcription — Translation — Primary structure

The array of amino-acids are detached from the tRNA and liked together to generate the primary structure of a protein (discussed in the next section). If this seems confusing, it's a hint of how difficult it was for early scientists to understand the flow of information.

While it seems logical to associate each amino-acid with the plus codon, the table of the Triplet Code, for historical reasons, lists the negative codes on the mRNA.

Sophisticated Information

Both computers and genetics evolved dramatically—one by human input from the mind, the other by input from the Logos. Computers evolved over 35 years from the 25k 8-bit Model 100 I first used as a word processor to the 16Gb 64-bit Mac Air I am now using. Bacterial prokaryote cells evolved over billions of years into the sophisticated eukaryotes of all higher life forms, including humans.

My current 64-bit computer manipulates arrays of 8 bytes at a time. One of those bytes is still the ASCII code generated as old by the keyboard. The other 7 bytes hold digital information about size, color, font, spacing, etc. which allows me to embed non-ASCII stuff such as photos, charts, videos while I can go crazy with typographical monstrosities such as

Letters to my aunt

This book, for example, is currently 193 pages long and in its full 64-bit glory takes up 89,917,173 bytes of memory. When saved as a *Text Only* file, however, where all the non-ASCII code is stripped away, it takes up only 404,000 bytes.

In a similar fashion, our eukaryote cells still use the triplet code established billions of years ago, but have added new layers of information alongside it. The triplet code info is contained in the *exon* stretches of DNA, these are separated by *introns* that are transcribed into mRNA but are stripped away by spliceosomes[1] in the nucleus. The remaining exons are spliced together to form the mRNA that is exported to the cytoplasm for translation into the primary structure of a protein, just as in bacteria. The nucleus exports *Text Only* transcripts, the introns remain in the nucleus and join in the activity of the CPU pool of RNA that runs the nucleus.

The untranslated introns can be very large, e.g., the human *GPIM* gene has 7 introns, the largest 4 having 32,638, 1,933, 1,657 and 1,283 base pairs respectively. For geneticists raised on bacterial genetics, this 'waste' made no sense, and the 95% non-translated DNA was labelled as parasitic junk and selfish genes[2]. This concept has now been discarded and non-translated RNA is currently stealing the spotlight from iconic exon DNA.

The pattern of splicing can change and different proteins created from the same DNA transcript. Anyone who has ever used boilerplate paragraphs to create different communications can probably relate. Even odder, the tem-

---

[1] https://www.nature.com/articles/nrm3742

[2] Dawkins, Richard R. (1976). *The Selfish Gene*. New York: Oxford University Press

plate and codon strands of DNA can switch roles—under proper guidance—and the codon info, not the template info, is expressed as protein. The bacterial simplicity has been replaced by eukaryote duplicity (in the eyes of the founders of genetics).

New, non-traditional RNAs			
Type	Abbr.	Function	Distribution
Antisense RNA	aRNA, asRNA	Transcriptional attenuation / mRNA degradation / mRNA stabilisation / Translation block	All organisms
Spliceosome RNA	snRNA	into excision, exon splicing	Eukaryotes
Signal recognition particle RNA	7SL RNA or SRP RNA	Membrane integration	All organisms
Transfer-messenger RNA	tmRNA	Rescuing stalled ribosomes	Bacteria
Cis-natural antisense transcript	cis-NAT	Gene regulation	
CRISPR RNA	crRNA	Resistance to parasites, by targeting their DNA	Bacteria and archaea
Long noncoding RNA	lncRNA	Regulation of gene transcription, epigenetic	Eukaryotes
MicroRNA	miRNA	Gene regulation	Most eukaryotes
Piwi-interacting RNA	piRNA	Transposon defense, maybe other functions	Most animals
Small interfering RNA	siRNA	Gene regulation	Most eukaryotes
Short hairpin RNA	shRNA	Gene regulation	Most eukaryotes
Trans-acting siRNA	tasiRNA	Gene regulation	Land plants
Repeat associated siRNA	rasiRNA	Type of piRNA; transposon defense	Drosophila
7SK RNA	7SK	negatively regulating CDK9/cyclin T complex	?
Enhancer RNA	eRNA	Gene regulation	?

IT'S A RNA WORLD

In the early days of genetics when DNA was unmasked, it was thought to be by its discoverers the "secret of life."[1] Nowadays, however, its role in genetics has shrunken to that of the hard drive in a computer. This, of course, is a crucial role as anyone who has ever experienced the loss of data in a crash will affirm. But it is a rather passive role as the long term storage of digital information and its passage down a lineage.

Just as the hard drive is useless, however, without the computer to use it; so DNA is useless without a set of RNAs to access it.

In the early days, the roles were reversed; RNA was the poor servant of DNA fulfilling a few traditional roles, albeit important ones, in the generation of protein.

More and more it is becoming the superstar of genetics, involved in the origin of life, having its information copied onto DNA (in violation of the Central Dogma of the Modern Synthesis[2] of Darwinism and genetics). These are just some of active roles played by RNA:

RNA was involved in the origin of living systems. RNA later shed some of its hydrophilic nature to become DNA. RNA controls the release and

---

[1] https://www.pbs.org/newshour/health/the-pub-where-the-secret-of-life-was-first-announced

[2] Crick, F.H.C. (1958). "On Protein Synthesis". In F.K. Sanders. Symposia of the Society for Experimental Biology, Number XII: The Biological Replication of Macromolecules. Cambridge University Press. pp. 138–163.

storage of digital information on DNA. RNA runs the command-and-control center of cells. RNA controls the generation of proteins. RNA controls the operation of the centrosome and transport in eukaryote cells.

These days, the RNA content of the nucleus seems to act more like the CPU at the heart of a computer manipulating digital content rather than a humble servant of DNA.

There is support for the idea that all the RNA in the nucleus—including the spliced out introns—act as the CPU of digital information processing:

> RNA is the computational engine of cell biology, developmental biology, brain function and perhaps even evolution itself. The complexity and interconnectedness of these systems should not be cause for concern but rather the motivation for exploring the vast unknown universe of RNA regulation, without which we will not understand biology.[1]

Analog Form

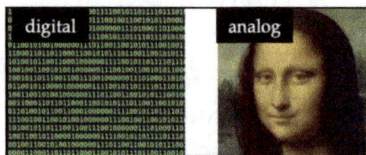

While digital information is eminently manipulable and dense, its analog expression is nonsensical. The digital information of the Mona Lisa photo is impossible to assimilate by our senses. The digital info my computer needs to illustrate the Mona Lisa is utterly unintelligible to us—it needs to be translated.

To summarize: In a simple computer system, a linear sequence of binary bits is translated into a linear sequence of letters and spaces. In living systems, a linear sequence of bibits is translated into a linear sequence of amino-acids.

In the earliest computers, the ASCII code called up a simple alphabetic form from a table of the monospaced COURIER font of a fixed 12-point size. Nowadays, the ASCII byte is embedded in many other bytes that determine font, color, size, etc., that feed into a mini program that generates a set a bezier points that determine length and curve at each point that together generate each letter. This allows me to type a wide range of character sizes just by change a scale parameter.

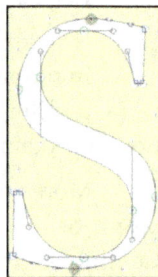

A similar development has occurred in sound and visuals so that having all three hours of *The Ten Commandments* on a 4.5" silver disc or video-chatting with a friend in Australia is nowadays considered unremarkable

---

[1] Morris, K. V. and Mattick, J. S., The rise of regulatory RNA. Nature Reviews Genetics 15:423-37 (2014), p. 432.

The relation between amino-acids and proteins is similar to the relation of bezier points and alphabetic forms used in computer typography. Each point has a snippet of information about relative length, curve, etc., that allows a computer to generate a character of any size quickly.

PROTEIN FOLDING

In a similar way, each amino-acid has a set of properties—size, water affinity, alpha coil or beta sheet tendency, etc.—that together generate the final folded state of the amino-acid chain. All this is intimately connected to the water structure surrounding the protein.

A patch, sometimes many patches, at the surface of the protein are not totally harmonious with the surrounding water, and find fulfillment by binding the substrate to this active site of an enzyme. This binding alters the whole protein structure and the substrate is modified then ejected for the process to repeat.

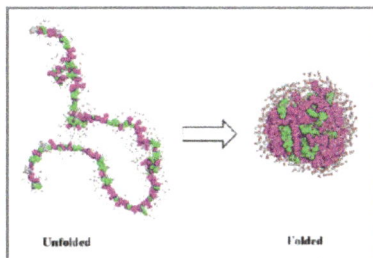

Unfolded   Folded

Cell communication

For most of the history of studying how cells in the body communicate and coordinate, it seemed that all the methods used were analog. One cell secreted a molecule which bound to a receptor on or in another cell and induced changes.

Simple examples of whole body effects are insulin—stimulating sugar uptake—and epinephrine (adrenaline)—promoting the fight-or-flight response. The paradigm of single cell-to-cell action is the axon of one neuron secreting a neurotransmitter—such as acetylcholine or dopamine)—which diffuses across a tiny gap (synapse) to bind with a receptor on a dendrite of another neuron and activating it.

As already noted, however, digital communication can be a lot easier and effective, as exemplified by the everyday difference between snail mail and email. So it seems strange that God does not seem to have used digital communication to communicate and coordinate cells in nature.

There are a few reasons to think that it is there, but not yet elucidated:

1. In this era of the Covid Pandemic we are very aware of the chaos a simple virus can cause. Infecting a cell, the virus releases a small amount of nucleic acid (RNA or DNA) that redirects the cell's metabolism into creating more virus particles. How is this possible? The simplest answer is

that a benign system exits to respond to digital input; the virus is just suborning this system.

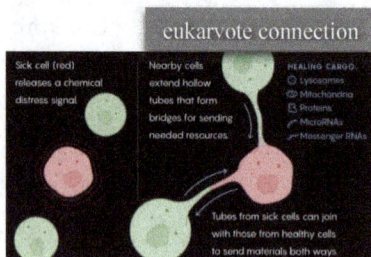
eukaryote connection

2. A recent discovery is that eukaryote cells have hollow filaments connecting them together:

> Long-overlooked "tunneling nanotubes" and other bridges between cells act as conduits for sharing RNA, proteins or even whole organelles.[1]

3. Bacteria transfer digital information to other bacteria, and as eukaryotes are direct descendants of them, there is no reason why they should not also do it. In bacteria, it is called conjugation, and the connection tube, a pilus. This is how, for instance, resistance to an antibiotic can spread through a population as the gene responsible is passed horizontally through a pilus from one to another. This is quite different from the vertical transmission of genes from one generation to the next.

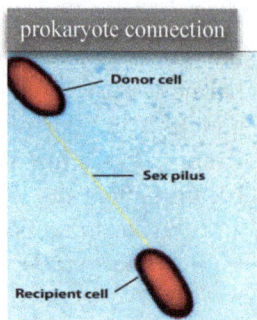
prokaryote connection

As the advantages of digital over analog communication are many, we can expect that this will emerge as a major way in which cells communicate and coordinate in the bodies of higher organisms.

Brain Function

In the early days off biochemistry's development, both proteins and nuclein (DNA, RNAs) were recognized as constituents of living organisms. Proteins were recognized as the master builders of life—as enzymes they ruled the chemical transformations of metabolism; as construction material they excelled as keratin, collagen, bone matrix; as movers and shakers of muscles, sphincters and peristalsis, and so on. Nuclein, on the other hand, was considered so simple that it was, perhaps, a structural component similar to the carbohydrate cellulose.

It was a great surprise, and Nobel worthy, when it was proved that genes were nuclein, not protein as was expected:

> Oswald Avery, Colin McLeod, and Maclyn McCarty showed that DNA (not proteins) can transform the properties of cells, clarifying the chemical nature of genes.[2]

---

1 Viviane Callier, *Cells Talk and Help One Another via Tiny Tube Networks*, Quanta Magazine, May 7, 2018

2 https://www.genome.gov/25520250/online-education-kit-1944-dna-is-transforming-principle

To summarize these disparate, yet connected, roles in living systems that work in harmony in the functioning of all living cells. A reflection of the dual characteristics of the Creator:

1.   Proteins are the master-manipulators of analog form—an external ability
2.   Nucleins are the master-manipulators of digital information—an internal ability.

It seems that a similar historical inversion is occurring in the study of brain function. It has bee long recognized that the brain of all animals—including humans—is constructed of two different classes of cells—the neurons and the glia.

Almost all studies of brain function have focused on the neurons: they make muscles contract; they process information from the optic nerve into a unified perception; they coordinate the activity of the heart, lungs, intestines, etc; they control the hearing and vocalizing of words; they wake us and put us to sleep; etc, etc. Neurons are the master-manipulators of the analog aspects of life.

Glia, on the other hand, were considered to be simple housekeepers, feeding and cleaning up after the neurons. But, like the history of the roles of protein and nuclein, it seems that a reversal is in progress:

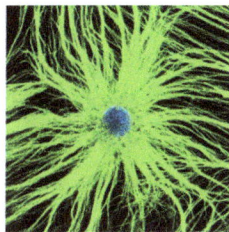

A Glia cell

> This [experimental result] is only one of many recent findings showing that glia, the motley collection of cells in the nervous system that aren't neurons, are far more important than researchers expected. Glia were long presumed to be housekeepers that only nourished, protected and swept up after the neurons, whose more obvious role of channeling electric signals through the brain and body kept them in the spotlight for centuries. But over the last couple of decades, research into glia has increased dramatically. ... It turns out that glia perform a staggering number of functions.
>
> They help process memories. ... while some communicate with neurons. Others are essential to brain development. Far from being mere valets to neurons, glia often take leading roles in protecting the brain's health and directing its development. Pick any question in the nervous system, and glial cells will be involved.[1]

In current thinking, the brain function of memory and learning involves the analog patterns of synapse strengths and weaknesses. It has already been established that glia control the formation and activity of synapses.

---

1 Elena Renken, *Glial Brain Cells, Long in Neurons' Shadow, Reveal Hidden Powers*, Quanta Magazine, January 27, 2020

Knowing the advantages of digital over analog information, we will not be going too far by supposing that eventually a new duality will emerge and the conclusion will be:

1. Neurons are the master-manipulators of analog form—an external ability

2. Glia are the master-manipulators of digital information—an internal ability.

Mind and Brain

The Unification Thought perspective on quantum physics suggests a solution to one of the great enigmas in classical physics: how does the intangible mind—which even devoutly materialistic scientists insist they possess—relate to and govern the substantial brain. Flex a finger; then try to explain exactly what you did to made it happen!

A clue to how this might happen is provided by the most consequential interaction on Earth. Without the interaction of light photons with the chlorophyll molecule there would be no plants; no food; no us. It is not hyperbole to say that this emergent property of photosynthesis from the Logos is the one on which all life on Earth is founded!

In the dark, the valence electrons from the central magnesium atom in chlorophyll are delocalized in the ground-state orbital around the central ring. When a photon of light is absorbed, the electron quantum jumps to an excited orbital that is localized on a part of the chlorophyll that is positioned at the top step of an electron cascade of cytochromes and other molecules.

The electron transfers to the top and tumbles down the cascade which captures its energy in proton-motive force that is used to generate ATP before returning to the ground state in the chlorophyll. Alternatively, the excited electron can transfer to another chlorophyll molecule where it is raised to an even higher level where is can activate hydrogen stripped from water leaving oxygen as a waste product.

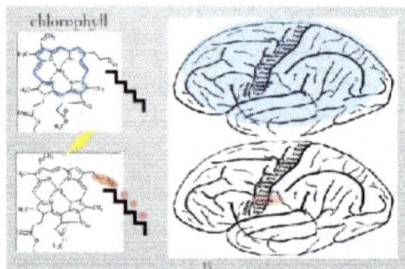

The ATP and activated hydrogen produced by the electron cascades are the used in the 'dark reaction' to drive the difficult chemical process of turning carbon dioxide into sugar, the basic food of all life.

Back to the brain and a speculation as to how a similar mechanism could explain the mind-body connection..

It is well-known from open-brain surgeries that there are distinct areas of the motor cortex of the brain that, when weakly stimulated by an electrode, will make fingers twitch, legs move, or lips pucker, etc.

First, we assume that while at rest, the wavefunction of the brain, aka the mind, is in a ground state that embraces the whole. Second, assume that the idea to twitch a finger cause the wavefunction—or the idea actually is the excited wavefunction—to concentrate on just that set of cells in the motor cortex that, when so triggered, send the set of impulses to the set of muscles that contract the ligaments that make the bones in finger twitch. The concentration of the wavefunction at that spot initiates a cascade that results in that set of cells firing, resulting in the finger twitching.

While it's a crude start to comprehending how the mind influences the body, it has at least the advantage over classical science which has no clue whatsoever to explain the mind-body connection.

The *élan vital* of Evolution

Many philosophers and some scientists have viewed Darwin's 'Random Variation and Natural Selection' as an improbable explanation of the evolution of life's sophistication. These historical perplexities are examined in a recent book[1] in which the author also notes the lack of alternative explanations. There are a few attempts, such as Bergson's *élan vital*, the medieval Vital Force, *vis viva*, biological *entelechy*, etc., but none have made an impression on modern science.

We shall look for an alternative explanation involving a previously unknown aspect of matter that is now fully-integrated into physics; the wavefunction. As noted, this is an abstract aspect to matter that can only be described by complex numbers—ones with a length and a rotation—unlike regular numbers that have just a length. In describing the wavefunction, the domain of length is 0 to 1 (impossible to certainty), while the domain of rotation is 0 to $\pm 180°$—the unit circle on the complex plane.

The wavefunction determines probability, so the interactions of two entities involve the overlapping of their wavefunctions, that add or multiply together following the laws of probability: The probability of A **and** B is $p(A) \times p(B)$; the probability of A **or** B is $p(A) + p(B)$.

The topic of adding and multiplying complex numbers leads almost automatically to discussing the Mandelbrot set (MS). This remarkable object is almost wholly within the domain of the wavefunction, the unit circle.

---

1 Neil Thomas, 2021, *Taking a leave of Darwin: a long time agnostic discovers the case for design.* Discovery Institute Press, Seattle.

The MS is generated by iteratively adding and multiplying a complex number. This can generate a series of bounded points on the complex plane, creating a Julia set with a distinct form. Such bounded numbers belong in MS. Alternatively, the series can move towards infinity and such unbounded numbers do not belong in MS.

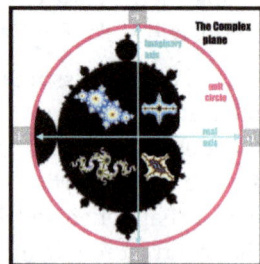

The diagram illustrates that part of MS that lies within unit circle, and four examples of the bounded Julia sets to be found within MS.

While the blending together of wavefunctions in the interactions of electrons and atomic nuclei are different to the creation of the MS, the form-making tendencies of complex numbers are clearly expressed in the intricate forms of the resultant orbitals that range from simple to intricate. These shapes contain single or paired electrons.

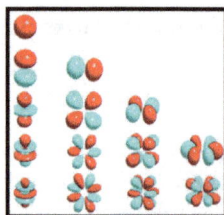

In Classical physics, where there is no concept of an internal wavefunction, regular numbers with magnitude suffice to encapsulate natural law. In the new physics where, as previously discussed, natural law acts on the internal wavefunction, it seems in inevitable that we need complex numbers to describe the natural laws of physics. This is true of Schrodinger's equation that describes the shape of atomic orbitals, and in the metric of complex Minkowski spacetime.

These two distinct forms with their distinct characteristics come from the forms in the natural law, the Logos discussed in an earlier chapter. If this holds for physics—the forms and character of atoms—and chemistry—the forms and characters of molecules—there is no reason to expect that it will not hold for all the levels of biochemistry and biology That every level of the Logos contains distinct probability forms that are expressed in nature— protein forms, nucleic acid forms, bacterial forms, cell forms and so on up the hierarchy until we reach the top, the form of a male and female human.

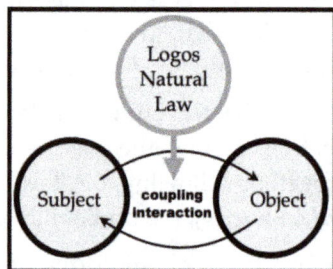

All these forms are similar in that they have a high probability when circumstances allow an interaction between entities to occur. The simple perquisites being: a) that the entities already exist and b) that their mutual histories have brought them into a configuration where they can interact.

A complex development can usually be broken down into a series of bilateral interaction; rather than three or more entities simultaneously interacting.

There is usually an initiating entity—the subject—and a responsive entity—the object, and their interaction will be guided by the appropriate form in natural law. The history of the interaction will move from less probable to more probable states, and the interacting entities will 'fall' into the highly probable form of the wavefunction.

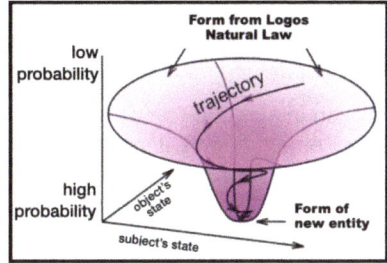

SCIENTIFIC *ÉLAN VITAL*

This is the scientific nature of the *élan vital* that, over time, moves up the hierarchy of gradually expressing forms at each level.

So, while we have no problem with the facts of evolution, we do disagree with Darwin's idea of the accumulation of fortunate random accidents. This is a quote[1]—lightly edited—to be found in a book by Richard Dawkins, a persuasive proponent of Darwin's point of view:

> We are walking archives of ancestral wisdom. Our bodies and minds are live monuments to our forebears' rare successes.... The human eye the brain, our instincts are legacies of [their] victories, embodiments of the cumulative experience of the past. And this biological inheritance has enabled us to build a new inheritance: a cultural ascent, the collective endowment of generations. (Helen Cronin, from The *Ant and the Peacock*)

Darwin and Dawkins agree that all this ancestral wisdom was pushed forward by accumulating wisdom by fortunate accident. The view presented here disagrees, the accumulation of ancestral wisdom was pulled forward by the probability forms inherent in natural law, the Logos: "Natural selection has a limited repertoire of potential forms from which to choose, and convergent evolution is the result."[2]

That the patterns for thriving are all in the Logos solves one of the major problems of modern Darwinism: "The modern synthesis is remarkably good at modeling the survival of the fittest, but not

---

[1] Richard Dawkins, *The Oxford book of modern science writing, Oxford University Press 2008, p.16*

2 George McGhee, Convergent Evolution: Limited Forms Most Beautiful (Cambridge, MA: MIT Press, 2011), 7.

good at modeling the arrival of the fittest."[1]

This is a thought by an avowed Darwinist—who, reading his brilliant book, emerges as a reluctant materialist—describing the course of evolution. He reluctantly views the course as arbitrary; which in the new view is a course laid out, and preplanned, by the Logos:

> An image that comes to mind is that of a surface of water slowly spreading over an irregular terrain. Fingerlets extend here and there, as local attractive forces battle with surface tension, until a minor breakthrough suddenly occurs in a given direction and all the pressure momentarily concentrates on a single rivulet. After that, groping soon resumes, sending out feelers until the next breakthrough.[2]

1 Scott Gilbert interview by John Whitfield, in "Biological Theory: Postmodern Evolution?" Nature 455 (September 17, 2008): 281–284, https://doi.org/ https://doi.org/10.1038/455281a.

2 Christian de Duve, 1995, *Vital Dust: Life as a Cosmic Imperative*, BasicBooks, NY, p. 93

8 • SEQUENCE OF EDENS

W hile both Science and the Bible agree that Creation happened in stages, science goes into a lot more detail and a lot more stages. Transition from a state that is devoid of a system to a state with many systems implies a sequence of origin events when the first of a system makes its appearance. All scientists and religionists agree on at least one point: the Universe started simple with an absence of sophisticated systems. The Bible states that in the first instant there was only light; while science asserts that this light contained a one-hundred-billionth 'impurity' of matter particles in it.

Contemporary science insists that evolutionary history from this simple origin is not teleological, that the doctrine of design and purpose in the material world is false. This atheistic dogma insists that every step in development in the history of life—from its abiotic beginnings to humans—must stand on its own merit in the struggle for survival. If is not useful now then it cannot survive to be useful later. The Logos of Unification Thought suggests otherwise—sometimes an aspect emerges that is not useful now but has a role to play in later developments.

Just Right!

The purposeful nature of the Logos—with the advent of the human capacity as its goal—is exemplified is the sequence of edens in the history of the universe. Our definition of a system-eden is the time and place where all the constituent subsystems are present, and the environmental circumstances are just right, for the very first of a system to appear in the history of the universe. This is an Origin Event. For example, the universe in both science and religion started with no hydrogen atoms. Nowadays there are plenty of them. So at some point in time the very first hydrogen atom emerged, followed by many others. The hydrogen-eden is the time and place when the very first hydrogen atom appeared in the universe—the Origin Event for hydrogen atoms.

While this is logically true, science has that almost immediately following this Origin event, there emerged tens, thousands, millions, billions, trillions, etc., of hydrogen atoms—the Origin event of the system was followed by similar multiplication of the system. As discussed earlier, it is here that we find the great divide between non-living and living systems: the Origin event and multiplication event of non-living systems are essentially the same; while they are fundamentally different in living systems. For all systems, however, the origin of the first one is the always similar, with the same two requirements: the subsystems interact under the direction of the Logos, and the environment must be just right for the new system to emerge and flourish—the eden for that system.

There are three stages in the emergence of an eden: First, a preparation period where the subsystems are all generated—the formation period of initiation; Second, a growth and development period where an environment suitable for their interaction emerges; Third, a perfection and completion stage where the subsystems come together and interact, and the new system emerges. This new system multiplies and develops into the formation stage for the next level of sophistication.

We shall now examine each eden in sequence. As modern science has well-established, creating physical stuff—given spacetime—is simple. All you do is pump energy into a small amount of spacetime and, voila! a speck of empty spacetime transmutes into equal amounts of matter and antimatter. While it's true that our universe is entirely[1] composed of matter, antimatter is also physical in that it has the exact same positive energy as matter does. It is every other character—charge, spin, q.color, topology, etc.—that is exactly the opposite.

So the very first eden we will need to deal with is 1) Origin of spacetime 2) Origin of matter, no antimatter.

SPACETIME EDEN

While the debate is still ongoing, about just what was going on at time zero in the Big Bang, there is a general agreement of how to describe it in published papers. A minute sample is reproduced here.[2]

The hieroglyphics here are advanced math, and the consensus is that math preceded the Big Bang and molded its form and development. As this field is constantly developing, I shall just give my thoughts here on the origin of spacetime and the matter/antimatter imbalance.

$$\left[\left(\frac{1}{R}\frac{dR}{dt}\right)^2 - \frac{8}{3}\pi G\rho\right]R^2 = 0 \qquad (19)$$
$$R^2 \neq 0 \Rightarrow \left(\frac{1}{R}\frac{dR}{dt}\right)^2 - \frac{8}{3}\pi G\rho = 0 \Rightarrow \left(\frac{1}{R}\frac{dR}{dt}\right)^2 = \frac{8}{3}\pi G\rho$$

The Logos in the Abstract Realm generated a speck of False Vacuum. There is a math description of this False Vacuum, a sample is shown here[3]. This would be of Planck- volume, temperature, size, energy density, etc. This Planck volume—packed with every possible particle—then entered a period

---

[1] there are some caveats, such as antineutrinos and PET scans.

[2] https://www.researchgate.net/publication/279446746_Graviton_and_cosmology_equations_before_the_Big_Bang

[3] https://www.cantorsparadise.com/quantum-tunneling-bubble-universes-and-the-end-of-the-world-749f31a9fa3b

of inflation, doubling every Planck moment. As described elsewhere, the separation of the quarks generated an immense energy that (almost) halted inflation, generating the Hot Big Bang and flipping the false vacuum into the regular vacuum of spacetime.

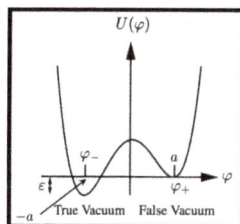

This is the origin event for spacetime and its eden was inflation. The Hot Big Bang was packed with equal amounts of matter and antimatter. Experimenters have already noted some subtle differences between matter and antimatter[1], and theory has it that at the Planck Temperature the math predicts that spacetime can transform into ±X particles, with odd properties such as ±4/3 charge, a mass greater than trillions of protons, and a quantum color as well as an anti-color.

As usual, most of these pairs annihilated into photons as the universe cooled. A very small percentage of them, however— perhaps 1 pair out of 100 billion—decayed as shown into three quarks and an electron. The R and B q-colors end up on the U quarks, while the two anti-colors combine into color G.

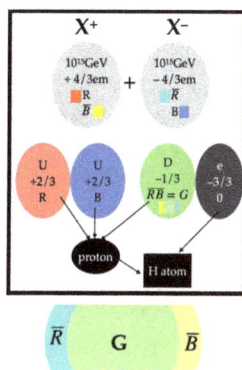

When all the matter and antimatter annihilated as the universe expanded and cooled (about 3 mins after time zero) the remaining quarks combined as a proton which, at a much later date, combined with the electron as a hydrogen atom. Each atom was accompanied by 100 billion photons (still around as the CMB).

ATOMS-EDEN

The Logos in the Abstract Realm generated a speck of false vacuum which inflated then flashed into the Hot Big-Bang of energy, slowing the rate of inflation, which cooled the ultra-hot universe as it expanded. (It is probable that the Dark Energy and Matter that nowadays make up 95% of the universe also emerged at this time, but as this is currently a known-unknown we will not speculate.)

This created the eden for the next step in the expression of the Logos, the emergence of the constituents of atoms out of which familiar matter is formed. This occurred about 13.5 billion years ago.[2]

The formation of the atomic eden was the dense, ultra-hot energy of the Hot Big-Bang quantized into every and all of the particles and antiparti-

---

[1] https://home.cern/science/physics/matter-antimatter-asymmetry-problem

[2] Tyson, Neil deGrasse and Donald Goldsmith (2004), *Origins: Fourteen Billion Years of Cosmic Evolution*, W. W. Norton & Co., pp. 84–5.

cles in the currently understood zoo of modern fundamental physics. The growth was the cooling of the still-expanding—albeit much slower—space-time just created. As the cooling advanced, heavy particles decayed into lighter particles, particles and antiparticles annihilated, until all that remained was a plethora of photons with a tiny percentage of matter particles.

The complexities of the Hot Big-Bang were essentially over after 3 minutes[1] resulting in a vast number of gamma ray photons with a 1 in a 100 billion 'contamination' of electrons, protons (75%) and helium nuclei (25%).

The Gamma ray photons, stretched by the continuing expansion and gradually losing energy resisting the Universe's expansion, are nowadays the low energy Cosmic Microwave Background (CMB) photons[2] that pervade all of space, while the regular matter is still ~75% hydrogen and 25% helium. This is the CS that provided the seeds of the next step, the eden that allowed for the emergence of atoms in the Universe.

The growth period was a hot plasma of gamma photons that prevented any protons and electrons from sticking together. For about $1/2$ million years after the Big-Bang, the universe was too hot for any atoms to form, but the expansion of the Universe sapped energy from the photons and they shrank to X-ray, then UV, then blue, red, IR photons and on down. While UV photons could disrupt atoms, the blue could not, and protons and electrons could stick together as hydrogen atoms, and electrons and helium nuclei could bind as helium atoms. This was the completion period, badly named recombination (since they had never been together).

The first, then a multitude of hydrogen and helium atoms appeared in the Universe. This set the stage for the next eden in which the elements essential for life—carbon, oxygen, nitrogen, sulphur, phosphorus, iron , etc.,—and the elements essential for the background for life—silicon, aluminum, etc.— were created.

METALS-EDEN

Unlike the chemists, astrophysicists classify all the elements in the Periodic Table—excepting the primordial hydrogen and helium—as 'metals'. Metals were not created in the Big-Bang eden, they emerged much later. Although the details are obscure, when the universe was less than a billion years old,[3] the primordial hydrogen/helium had gravitationally coalesced in to galaxies and the first generation of stars.

---

[1] Steven Weinberg, *The First Three Minutes: A Modern View of the Origin of the Universe*, Basic Books 1977

[2] http://planck.cf.ac.uk/science/cmb

[3] http://firstgalaxies.org/explore.html

"Theorists predict that the clouds of gas in the early universe would have remained relatively warm from the Big-Bang and so would resist condensing down to form stars. Mixing in a small amount of heavier elements helps gas clouds cool, because those elements are easier to ionize and so shed heat as radiation. But those heavy elements hadn't yet formed in the early universe, so stars grew to enormous sizes—hundreds or even a thousand times as big as our sun—before their cores were dense enough to spark fusion. Once they did get started, they burned fast and hot, emitting lots of ultraviolet light and burning out in a few million years"[1]

It is in the cores of these 1st-generation stars that hydrogen fused to helium. It is the energy released by this fusion that fuels the stars on the Main Sequence.[2] Leaving the Main Sequence as the hydrogen was depleted, the stars entered the final, and much shorter, stage of a red giant when helium fused to carbon (as discussed earlier) and oxygen, and so on up to iron. This is the development in the creation of the non-primordial elements in the Periodic Table.

The formation of iron, however, is the death knell for a star as fusion can no longer generate energy. The death throes of a giant star is a supernova—brighter than an entire galaxy of billions of stars—that scatters the metals formed within the star into the interstellar medium to participate in the formation of the 2nd- and 3rd-generation stars. This hyper-explosion is so energetic that it forces nuclei together to create all the elements more massive than iron.

This is the completion stage where 'metals' appear in the interstellar gas out of which new stars form. The 2nd generation were less massive, and the 3rd generation—of which our Sun is a member—smaller still and with lifetimes in the billions of years. Sufficient metals were in the gas that condensed into the solar system to allow rocky planets—such as the Earth—to form. This was the creation of the eden for life to emerge, and our focus shifts from the Universe as a whole to this specific planet.

Simple Life edens

About 4.5 billion years ago, our Sun condensed along with its suite of planets. The details of the formation of the Earth, the eden of life, are still a matter of debate:

"The question of the origin of the solar system is one that has been a source of speculation for over a hundred years; but, in spite of the attention that has been devoted to it, no really satisfactory answer has yet been

---

[1] http://www.sciencemag.org/news/2015/06/astronomers-spot-first-generation-stars-made-big-bang

[2] http://www.atnf.csiro.au/outreach/education/senior/astrophysics/stellarevolution_mainsequence.html

obtained. There are at present three principal hypotheses that appear to contain a large element of truth, as measured by the closeness of the approximation of their consequences to the facts of the present state of the system, but none of them is wholly satisfactory."[1]

One thing that is very clear, however, that it was the presence of the Moon that made the Earth the eden for the emergence of living systems; for without it the Earth would be quite different—less benign and more hostile.[1]

Evidence is accumulating that the Moon formed when a Mars-sized planetoid collided with the early earth. "At the time Earth formed 4.5 billion years ago, other smaller planetary bodies were also growing. One of these hit earth late in Earth's growth process, blowing out rocky debris. A fraction of that debris went into orbit around the Earth and aggregated into the Moon."[2]

There is another aspect of the Earth that makes it conducive to life, that is the plate tectonics that slowly reconfigures the Earth's surface, creating mountains and volcanoes. At the plate boundaries, heat and material from deep inside escape in a variety of "smokers" that many suggest were crucial to life's origin, as discussed in the following section. There is no consensus as to why Earth has tectonics while the similar planets of Venus and Mars do not:

"When and how plate tectonics started is a key question among geologists. Some researchers think it started more than 4 billion years ago, and others say it started only about 1 billion years ago. That's a big range, and the uncertainty stems from the fact that it's simply hard to find well-preserved ancient rocks."[3]

One possibility is that the immense collision that splashed off the Moon also fragmented the surface of the Earth, creating the tectonic plates that slowly migrate across the globe. Without the Moon, history might have been akin to Venus, where the lack of surface fractures allowed a buildup of internal pressure that was released in a global vulcanism that obliterated the entire surface:

"Venus underwent a global resurfacing event 300–600 [million years] ago, the cause and nature of which remains uncertain. The present-day surface heat flux on Venus is about half the likely radiogenic heat generation rate, which suggests that Venus has been heating up since the resurfacing event."[4]

Earth's tumultuous development period, the Hadean time (4.6 to 4 billion years ago) ended as the planet cooled, the Earth's crust formed, the

---

[1] https://nineplanets.org/questions/what-would-happen-if-there-was-no-moon/

[2] https://www.psi.edu/epo/moon/moon.html

[3] https://www.livescience.com/31570-plate-tectonics-began.html

[4] F. Nimmo and D. McKenzie, Annu. Rev. Earth Planet. Sci. 1998. 26:23–51

oceans condensed out, and the establishment of the environment just right for living systems to emerge.

Last Universal Common Ancestor

The tumultuous birth of the Earth/Moon system ended ~4,400 million years ago as the oceans were established. Scientists assumed, at first, that the transition from the chemical era to the biochemical era must have been unlikely and taken eons to occur. The Logos and the Wavefunction, however, had quite different concepts of what was probable or improbable! Thus it came a quite a shock when signs of living systems were found in rocks that were 4,300 million years old.

There are many theories of the events that changed non-living chemicals into an assembly of them that was alive. One theory seeks to explain why all life uses acid gradients to power chemical interactions—the universal proton motive force of photosynthesis and respiration—by exploring this aspect of mineral chemistry:

Scientists exploring the inorganic origins of life have used hydrothermal minerals to catalyze the difficult reaction between H_2 and CO_2, forming acetate and pyruvate, critical constituents of the reverse Krebs cycle. Pleasingly, these mineral catalysts include greigite, an iron sulfide with a similar basic structure to the iron–sulfur clusters in ferredoxin, the protein that still catalyses the two most difficult steps in cells today. To my mind at least, it's not coincidental that the Earth provides CO_2 and H_2 in buckets, along with the iron–sulfur catalysts that facilitate their reaction to form carboxylic acids, the same reverse Krebs cycle intermediates that are still at the heart of metabolism today.[1]

While there is still much debate as to the sequence and location of the events that organized simple chemicals into a living organism[2]—life being a quality inherited form the Logos—the end result was the last universal common ancestor (LUCA) from which all past and extant life is descended.[3]

One aspect of the origin of life question that it is very clear is that, whatever the process was, it ended with a single entity that was the ancestor of all known life, the Last Universal Common Ancestor, LUCA. This is because all of life—from bacteria to trees to humans—have so much in common, including these five features:

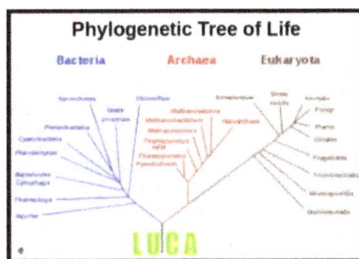

Phylogenetic Tree of Life

---

[1] Lane, Nick. *Transformer: The Deep Chemistry of Life and Death* (p. 121). W. W. Norton & Company.

[2] An excellent overview of this area of research in Christian de Duve's *Vital Dust* pp, 15-51

[3] https://www.nytimes.com/2016/07/26/science/last-universal-ancestor.html

1.   DNA ➜ mRNA ➜ rRNA ➜ tRNA ➜ protein ➜ metabolic activity,

2.   Universal Triplet Code specifying the 20 ubiquitous amino acids,

3.   Chirality: L-amino acids, D-nucleotides used in all of life,

4.   ATP energy manipulation, NAD hydrogen manipulation,

5.   Proton Motive Force to create ATP from AMP.

As chemical energy was required to drive this development, a well-accepted theory is that the edens for these events were the volcanic hydrothermal vents, the black and white smokers found today along the tectonic plate boundaries spewing high-energy compounds into the cool ocean:

> "Hydrothermal vents — where heated, mineral-laden seawater spews from cracks in the ocean crust — created a gradient in positively charged protons that served as a "battery" to fuel the creation of organic molecules and proto-cells."[1]

The LUCA is estimated to have lived some 3.5 to 3.8 billion years ago[2] and had the basic aspects shared by all current life. The LUCA lineage was the completion stage of basic life, and diverged over time to generate the three great domains of life: the bacteria, the archaea and the eukaryotes—elucidated by ribosomal structure.[3]

These days, the archaea flourish in conditions similar to those found on the early Earth—hot, acidic, salty, sulfurous, etc.—considered hostile and avoided by all other living organisms. The bacteria are everywhere else and essentially unchanged. In the 1500 million years following LUCA they transformed the Earth by generating an oxygen atmosphere as well as creating an eden for the eukaryotes to emerge, setting the stage for the expression of the sophisticated levels of the Logos as plants and animals. The diagram illustrates roughly the current consensus as to the Earth's history using a scale of millions of years before present.

0	HUMANS
	FLOWERS
	MAMMALS
	DINOSAURS
-500	LAND LIFE
	CAMBRIAN EXPLOSION
-1000	SEXUAL REPRODUCTION
-1500	MULTICELLULAR ORGANISMS
-2000	ORIGIN OF EUKARYOTES
-2500	OXYGEN ATMOSPHERE
-3000	
-3500	FIRST PHOTOSYNTHESIS
-4000	LUCA
	EARLIEST SIGNS OF LIFE
-4500	FORMATION OF OCEANS
	FOMATION OF EARTH

Both the bacteria—technically the eubacteria—and the archaea are prokaryotes, they are characterized by a lack of internal organelles, espe-

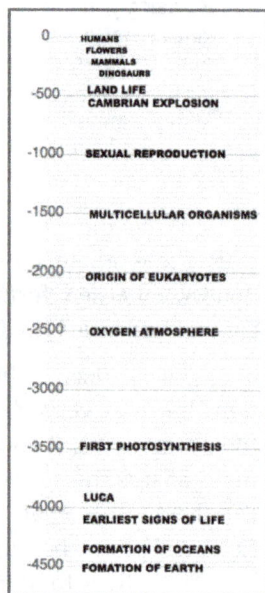

---

[1] https://www.livescience.com/26173-hydrothermal-vent-life-origins.html

[2] Theobald DL (May 2010). "Formal test of the theory of universal common ancestry". Nature. 465 (7295): 219–22.

[3] https://en.wikipedia.org/wiki/Last_universal_common_ancestor

cially the nucleus. The main metabolic challenge facing living organisms is adding hydrogen to carbon dioxide to generate carbohydrates which are the basis for every other type of life's molecules. The challenge is to find a source of hydrogen, and to obtain sufficient energy to drive the reaction. While early metabolism probably reduced CO_2 to sugar using hydrogen sulphide as a source of hydrogen (laying down the ancient beds of sulphur mined today) and natural chemical energy in molecules such as thioesters.

Eventually the Logos guided the development of a metabolic pathway that used the ubiquitous water molecules as a source of hydrogen and the capture of red and blue light energy to drive the reaction. The discovery of photosynthesis, which uses the energy of the abundant light quanta emitted by the Sun to separate water into hydrogen and oxygen, liberated bacteria from their dark origins to freely populate the oceans.

The oxygen liberated by the photosynthetic bacteria was first absorbed by the ocean's soluble ferrous iron which converted to insoluble ferric iron which precipitated out as the immense banded iron deposits which are the source of modern-day iron ore:

"We know there was some free oxygen in the atmosphere by 2.3 or 2.4 billion years ago, but it took until around 2 billion years ago, after 700,000,000 years of work by the cyanobacteria, for there to be enough oxygen in the atmosphere to think of it as relatively oxygen rich. For a while, a few hundred million years, the highly reactive oxygen given off by photosynthetic organisms probably combined with iron dissolved in the early oceans, so oxygen didn't accumulate in the atmosphere. It produced thick iron oxide deposits like those in Minnesota..."[1]

Eukaryote eden

While many photosynthetic bacteria floated free in the primordial ocean, some remained attached after division which lead to the formation of stromatolites. These are layered bacterial structures that occur widely in the fossil record of the Precambrian, > 600 Million years BP, but are rare today.

"Modern stromatolites are mostly found in hyper-saline lakes and marine lagoons where extreme conditions due to high saline levels prevent animal grazing."[2]

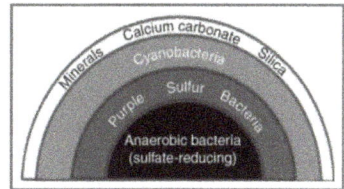

The layering of the stromatolites has the primary producers photosynthesizing in the top layer. Their dead bodies

---

[1] http://historyoftheearthcalendar.blogspot.com/2014/01/january-19-oxygen-crisis-2-billion.html

[2] https://en.wikipedia.org/wiki/Stromatolite

feed the lower layers. It as probably in such a protected environment that some prokaryotes could shed their protective coating and adopt phagocytosis—feeding by engulfing food with the now-freed flexible membrane. Modern day freely-living amoeba and the macrophages in our blood still use this ancient method to respectively feed and clear the blood of bacteria and debris. Rather than an outer coat, an internal network of protein filaments was developed to control the cells structure:

> "The cytoskeleton is composed of three distinct elements: actin microfilaments, microtubules and intermediate filaments. The actin cytoskeleton is thought to provide protrusive and contractile forces, and microtubules to form a polarized network allowing organelle and protein movement throughout the cell. Intermediate filaments are generally considered the most rigid component, responsible for the maintenance of the overall cell shape."[1]

Other steps in this stromatolite eden were the enclosure of the DNA in a bi-lipid membrane—the origin of the nucleus—the symbiosis with an ingested but not digested prokaryote that could utilize oxygen efficiently—the origin of the mitochondria—and for a later lineage, a similar symbiotic relationship with a photosynthesizing prokaryote—the origin of the chloroplasts. These two examples are of endosymbiosis:

"The endosymbiotic origin of mitochondria and chloroplasts is widely believed because of the many similarities between prokaryotes and these organelles:

1. Mitochondria and chloroplasts are similar in size and shape to prokaryotes
2. They have their own DNA that lack histone proteins, are circular, and are attached to the inner membrane as is the DNA of prokaryotes
3. Their ribosomes are more similar in size to prokaryotic ribosomes
4. They divide by fission, not mitosis.
5. Mitochondria arise from preexisting mitochondria; chloroplasts arise from preexisting chloroplasts (they are not manufactured through the direction of nuclear genes).
6. Their outer membrane would have been synthesized by the original "host" cell and used to engulf the endosymbiotic bacteria that became the organelles. Their outer membrane has structural and chemical similarities to the eukaryote cell membrane."[2]

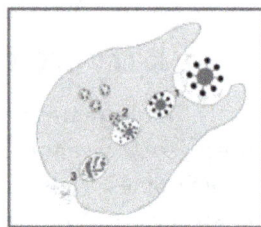

[1] https://www.ncbi.nlm.nih.gov/pubmed/15180824

[2] https://www2.gwu.edu/~darwin/BiSc151/Eukaryotes/Eukaryotes.html

The end result of this long development growth stage was the eukaryote cell, a lineage that later developed into all the non-bacterial forms of life: fungi, plants and animals.

ORIGINS OF SEX

Asexual reproduction is simple; the DNA is duplicated and the cell divides with each half getting a set of DNA:

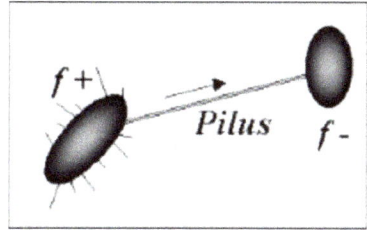

Though bacteria are predominantly asexual, the genetic information in their genomes can be expanded and modified through mechanisms that introduce DNA from outside sources. Bacterial sex differs from that of eukaryotes in that it is unidirectional and does not involve gamete fusion or reproduction.[1]

Eukaryote sex is mutual, involves gamete fusion, and is central to reproduction while the passing on of mitochondria and chloroplasts is uniparential, solely through the maternal lineage:

Sexual reproduction is a nearly universal feature of eukaryotic organisms. Given its ubiquity and shared core features, sex is thought to have arisen once in the last common ancestor to all eukaryotes. Using the perspectives of molecular genetics and cell biology, we consider documented and hypothetical scenarios for the [origin] and evolution of meiosis, fertilization, sex determination, uniparental inheritance of organelle genomes, and speciation.[2]

One of the few extant eukaryotes that do not have endosymbionts— either mitochondria or chloroplasts —is Giardia that is probably a remnant of the earliest eukaryotes. Like many eukaryotes it is diploid—having two sets of essentially identical chromosomes. Unlike the others, however, these inhabit two separate haploid nuclei.

The advantages of the diploid over the haploid state is still a matter of debate[3] and many simple plants spend much of their life cycle in the haploid state. In all higher organisms, however, the haploid state is transitory while it is the diploid state that is predominant.

Multicellular edens

In the discussion so far, we have been dealing with single cells or colonies of single cells such as the stromatolites. Such colonies are not considered multicellular as all the members are identical clones. While the his-

---

[1] http://www.sciencedirect.com/science/article/pii/S0960982206019725

[2] http://cshperspectives.cshlp.org/content/6/3/a016154.full

[3] http://www.genetics.org/content/156/2/893

tory of life's development of Logos-related structures so far has covered over three billion years, all this occurred in the world ocean. The challenges of populating dry land were such that it only multicellular organisms—characterized by cells differentiated into a variety of forms dealing with different aspects of the challenge—that could accomplish this feat. The eukaryote lineages of autotrophs and heterotrophs diverged as multicellular forms were explored:

> "Multicellular eukaryotic forms of life probably arose initially from small clones of cells that remained associated after their production, by successive divisions from a single parental cell. The cells were held together either by intracellular connections or a shared external wall or shell. Roughly speaking, the former mechanism led to animals and the latter to plants and fungi"[1]

In this era, the oceans would have hosted a plethora of photosynthetic and scavenger eubacteria—the archaea thriving only in the remnants of the hadean era—and photosynthetic and scavenger eukaryotes, the protists common in the ocean and ponds to this day. The photosynthetic cyanobacteria—in an earlier time called the blue-green algae for their habit of sticking together in long chains—were probably the first to explore the advantages of creating a colony of clones.

SEAWEED EDEN

It was probably in the tidal littoral zone—where the sea first covered then exposed—that plants discovered the advantages of sticking to one place. At one end of the clonal chain, cells secreted chemicals that fixed them to the rocks—the holdfast of seaweeds—and others created buoyant floats that lifted the fronds to the light—while the remaining cells spread out and focused on photosynthetic growth.

In this littoral eden the seaweeds perfected themselves and have changed little since. As this mode was eminently suited to the ocean, this was as sophisticated as multicellular eukaryotic plant life developed in the ocean; an example of the truism: "If it ain't broke, don't fix it."

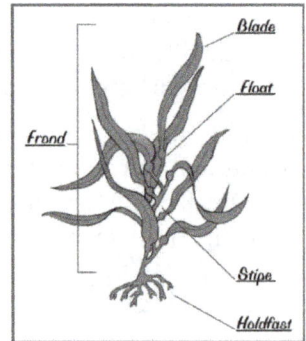

The exploitation of the heretofore barren dry land began with the simple mosses that often spend half of their lifecycle in the haploid state—

---

[1] Christian de Dove, *Vital Dust*, BasicBooks 1995, p.176

the diploid and haploid stages looking very similar to each other. This is called alternation of generations:

"In this life cycle, a haploid organism (the gametophyte) produces gametes by mitosis. These gametes fuse in a fertilization event, creating a diploid zygote. This diploid zygote divides (typically) mitotically to produce a multicellular diploid offspring. This organism produces haploid spores by meiosis. These spores develop, dividing mitotically to produce the next multicellular haploid generation. So, the ploidy level 'alternates' across the generations. In green algae, the sporophyte and gametophyte stages may be 'isomorphic' (look the same)…"[1]

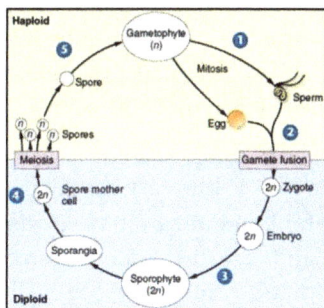

Land plant edens

The advance of plant life onto dry land was progressive, following a gradient in the marshy land near water, as new skills were inherited from the Logos.

MOSS EDEN

The first skill was to avoid desiccation by developing a waxy covering with openings—primitive stomata—for CO_2 to enter and O_2 to depart. Not being blown away from water was also a needed skill provided by roots. Finally, reproduction without cells traversing open water was necessary, and tough coatings allowed haploid spores to disseminate with the wind and germinate when water was found. The haploid plants produced motile male and immobile female gametes which could meet, fuze, grow and generate haploid spores.

A green carpet extended from open water onto the previously barren land. These days there are some 15,000 species of moss exploiting a plethora of ecological niches.

While greatly different in its outer reaches, the metabolism of all living systems is the same at the central core where carbohydrates activated by phosphate are manipulated. Plants create carbohydrates by capturing energy from sunlight; fungi and animals obtain energy by breaking down carbohydrates. Their name suggests that carbon and water are united, but this is a misnomer: plants use energy to strip hydrogen from water (liberating oxygen) and add it to carbon dioxide. Animals strip the hydrogen and add it oxygen (recreating water) liberating energy and carbon dioxide.

---

[1] http://facweb.furman.edu/~wworthen/bio111/plant1.htm

In plants, light energy is used to generate molecules of ATP and NADPH (activated hydrogen)—the light-dependent reactions. To make a molecule of glucose in the light-independent reactions, six CO_2 molecules are hydrogenated by twelve NADPH driven by the energy liberated by the breakdown of eighteen ATP in a complex cycle of transformations known as the Calvin Cycle.[1]

While carbohydrates are the feedstock for generating amino-acids, fats, nucleotides and all the other molecules that plants manufacture, glucose itself is used directly to generate two important macromolecules, starch and cellulose. The great difference between these two molecules is the way the glucose monomers are linked together. The seemingly insignificant difference, as illustrated, determines two very different properties inherited from the Logos; starch being the edible stuff of potatoes, flour and white rice—cellulose being the inedible-to-animals[2] tough fibers of grass, leaves and wood.

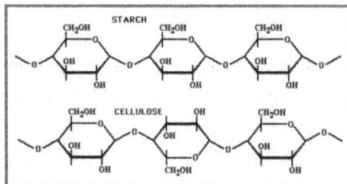

The toughness of cellulose was utilized in the next step of plant evolution as the process of vascularization was learnt from the Logos. Moving away from free water was only possible for plants with deep roots that could tap water underground, a location that precluded photosynthesis. Connecting the colorless roots and the green leaves were connecting stems, strengthened by cellulose with uni-directional xylem tubes transporting water and minerals upwards, and phloem tubes transporting photosynthetic products downwards.

Surprisingly, the exact mechanism of sugar transport in the phloem is not known, but it is certainly far too fast to be simple diffusion. The main mechanism is thought to be the mass flow of fluid up the xylem and down the phloem, carrying dissolved solutes with it. Plants don't have hearts, so the mass flow is driven by a combination of active transport (energy from ATP) and evaporation (energy from the sun).[3]

The vascular plants, unlike the mosses, developed a diploid mature phase while the haploid stage was reduced to a short stage, often underground, of generating gametes that fused into a diploid zygote that developed into the mature plant.

---

[1] https://www.khanacademy.org/science/biology/photosynthesis-in-plants/the-calvin-cycle-reactions/a/calvin-cycle

[2] Herbivorous animals harbor symbiotic bacteria that can break down the cellulose for their hosts. Termites do the same thing with wood.

[3] http://www.biologymad.com/master.html?http://www.biologymad.com/planttransport/planttransport.htm

About 400 million years ago, these developments allowed plants to cover the land in green and soil. With the discovery of lignin, a tough multi-linked polymer that strengthened cellulose, solid trunks allowed trees to grow to 40 feet or more.

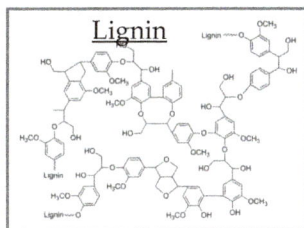
Lignin

As it took the bacteria and fungi time to learn how to digest lignin, the dead remains of these plants were not degraded but eventually fossilized creating the massive beds of coal found worldwide. In consequence, the period between 360 and 286 million years ago is called the Carboniferous era.

CONIFER EDEN

During the Carboniferous era (359-299 mya), the tectonic plates had been slowly been rearranging the continents. This productive Carboniferous era drew to a close as, when all the continents came together as one, called Pangaea, its center turned into a vast, arid desert. Much of this supercontinent was centered on the South Pole covered with a huge deep ice sheet, and the ocean level dropped

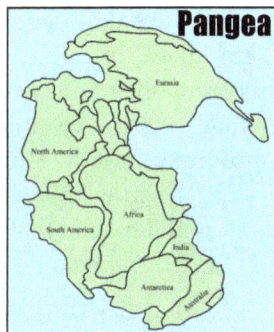
Pangea

In the same period, the massive volcanic eruptions in what is now Siberia generated great dust clouds, cooling the Earth globally, and the Earth entered the bleakest ice age. This was the start of the Permian Age and a wave of extinctions called the Permian Extinction:

A series of extinction pulses that contributed to the greatest mass extinction in Earth's history.... The Permian extinction was characterized by the elimination of over 95 percent of marine and 70 percent of terrestrial species. In addition, over half of all taxonomic families present at the time disappeared. This event ranks first in severity of the five major extinction episodes that span geologic time.[1]

Plants reacted to this challenge by reproducing with resistant seeds, rather than venerable spores. Extant species developed during this challenging period are the ginkgoes, pines, spruces, redwoods and other conifers. This was the harsh eden in which the superfamily of gymnosperms developed which, to this day, thrive in subarctic climates close to the poles or in the higher elevations of mountain ranges.

---

[1] https://www.britannica.com/science/Permian-extinction

FRUIT & FLOWER EDEN

About 100 million years ago, Pangaea had moved northwards and divided into the familiar continents and were moving into the configuration of today's globe. Fossil evidence indicates that flowering plants first appeared about 125 mya, and were rapidly diversifying by 100 mya. The flowers were a sign of the symbiosis between flowers—that rewarded with sugars—and the insects that carried pollen away to other flowers.

Along with this came the development of fruits, enveloping the seeds; flowers and fruit being the defining characteristic of the other superfamily of angiosperms, basically all familiar plants that are not conifers.

The angiosperms developed a process unique to plants: double fertilization. One pollen arrives and sends a male haploid gamete to unite with a haploid female gamete to generate the zygote seed that will develop into a new plant. A second pollen unites with a diploid cell in the female ovary—creating a triploid cell—which multiplies into the fruit which surrounds the seed. The fruit initiated another mutual relation with animals that, attracted and feeding on the fruit, spread the seeds far and wide.

The diagram illustrates the history of the major innovations learnt from the Logos by the ancestral eukaryote.

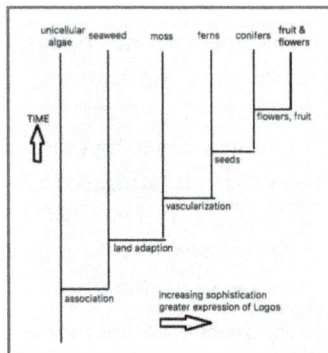

Along with the plant invasion of dry land came an army of scavengers related to yeast and lacking chlorophyll. Unlike animals, they were not mobile, reproduced by spores and relied on extracellular digestion. This was the origin of the fungi now encompassing over 200,000 different species, including the delicious mushrooms enjoyed by multitudes.

As both herbivorous and carnivorous animals ultimately depend on plants to generate food from air, water and sunlight, the invasion of land by animals of necessity followed that of the plants. While the plants never really progressed beyond seaweed in the ocean, the progressive sophistication of the animals took many steps in the ocean before attempting to follow the plants onto dry land.

Ocean Animals Edens

It is probably due to the complexity involved in the transition from a single-cell zygote to an intricate adult, but both animal evolution and development are remarkably conservative. Whatever steps occurred in the ancestral lineage are repeated, at least ephemerally, in the steps of development from a single-cell zygote to multi-cell adult. This is summarized as:

'Ontogeny recapitulates phylogeny' a catchy phrase coined by Ernst Haeckel, a 19th century German biologist and philosopher to mean that the development of an organism (ontogeny) expresses all the intermediate forms of its ancestors throughout evolution (phylogeny).[1]

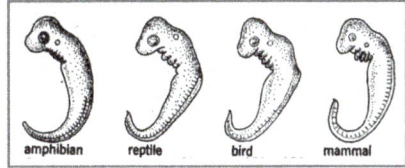

amphibian   reptile   bird   mammal

The evolution and development of the human form are central to the Logos, and all body plans are variations on this basic pattern. In fact, in the earliest stages of all animal development—the embryo—it is almost impossible to tell who will develop into an amphibian, reptile, bird or mammal.

BODY-PLAN EDEN

While survivors of the earliest days still remain today, it must be remembered that their ancestral lineages have hundreds of millions of years to develop and improve on their basic plan. With this in mind, an example of a sophisticated single celled heterotroph is the choanoflagellates, here illustrated, that are found all over the word in water environments where the play a key role in the microbial population:

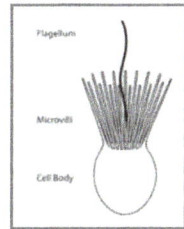

Flagellum
Microvilli
Cell Body

"In addition to their critical ecological roles, choanoflagellates are of particular interest to evolutionary biologists studying the origins of multicellularity in animals. As one of the closest living relatives of animals, choanoflagellates serve as a useful model for reconstructions of the last unicellular ancestor of animals."[2]

Such cells joined into spherical associations that combined cooperative propulsion and feeding, and with the development of internal channels became the sponges of our era. Invagination created a cavity for more entrapment and leisurely digestion, and this eventually developed into a primitive gut lined with endothelial cells, with ectoderm cells on the outside. This two-layers of cells is the diploblast pattern found in simple animals such as the hydra ubiquitous in ponds. This reproduces by budding off miniature copies of itself and the single opening serves as both a mouth to introduce prey to the gut cavity and an anus to rid it of undigested detritus.

In the development of the human zygote, ontogeny recapitulates phylogeny as the single cell first multiplies forming a hollow sphere of cells

---

[1] http://2000clicks.com/graeme/langwisdomsayingontogenyrecapitulatesphylogeny.htm

[2] http://tolweb.org/Choanoflagellates/2375

which then invaginates in gastrulation to form a double layered sphere, the gastrula, with a single opening, the blastopore. The ectoderm will develop into the skin and nervous system et al. while the endoderm into the gut lining, the liver and the lungs et al.

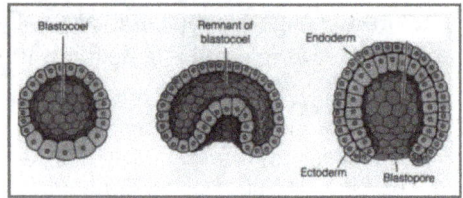

There were two major advances that followed, and it is currently unknown in which order they occurred historically:

1.   A second opening developed, the first remained a mouth into which prey were ingested, while the second became an anus out of which undigested detritus was expelled.

2.   A third layer of cells, the mesoderm, developed between the ectoderm and the endoderm surrounding an internal space called the coelom.

This is the triploblastic pattern that is the basis for most animal life. It is an important step in the development of the embryo:

The mesoderm is the middle of the three germ layers, or masses of cells (lying between the ectoderm and endoderm), which appears early in the development of an animal embryo. In vertebrates it subsequently gives rise to muscle, connective tissue, cartilage, bone, notochord, blood, bone marrow, lymphoid tissue, and to the epithelia (surface, or lining, tissues) of blood vessels, lymphatic vessels, body cavities, kidneys, ureters, gonads (sex organs), genital ducts, adrenal cortex, and certain other tissues.[1]

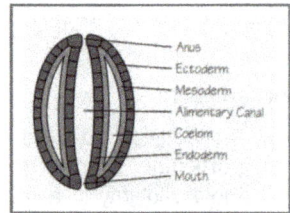

Examples of descendants of such primitive triploblasts are the flatworms, nematodes and the parasitic tapeworm and ascaris. Currently, the best understood animal is the nematode *Caenorhabditis eleegans*:

Around the world thousands of scientists are working full time investigating the biology of C. elegans. Between October 1994 and January 1995, 73 scientific articles about this creature appeared in international science journals. Currently an international consortium of laboratories is collaborating on a project to sequence the entire 100,000,000 bases of DNA of the C. elegans genome.... C. elegans is about as primitive an organism that exists which nonetheless shares many of the essential biological characteristics that are central problems of human biology. The worm is conceived as a single cell which undergoes a complex process of development, starting with

---

[1] https://www.britannica.com/science/mesoderm

embryonic cleavage, proceeding through morphogenesis and growth to the adult....All 959 somatic cells of its transparent body are visible with a microscope, and its average life span is a mere 2-3 weeks. Thus C. elegans provides the researcher with the ideal compromise between complexity and tractability.[1]

It was the triploblastic pattern that, under the guidance of the Logos, was the basis for the proliferation of animal body patterns that appeared in the Cambrian explosion of life 600-520 million years ago, whose fossils deposited in Canada were the topic of *Wonderful Life: The Burgess Shale and the Nature of History* by Stephen Jay Gould.

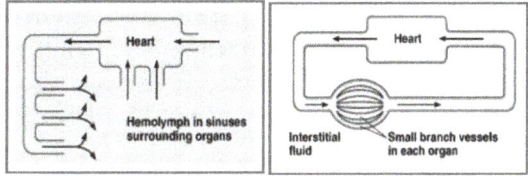

It was during this period of exploration of body patterns that a solution to the challenge of oxygen transfer and food distribution to internal tissues was discovered in the development of the *milieu interieur*, the bodily fluids regarded as an internal environment in which the cells of the body are nourished and maintained in a state of equilibrium. This involved learning three things from the Logos:

1. Simple gills, folded skin enabling efficient oxygen transfer from the ocean water.

2. The development of oxygen-carrying molecules such as the red hemoglobin, using iron, and blue hemocyanin, using copper.

3. Hearts to move the fluid about. In the earliest form this was just a thickening of the tube connecting to the coelom, an open circulation of hemolymph, and later the development of closed tubes with blood inside and lymph outside.

Open circulatory systems are still used by crustaceans, insects, mollusks and other invertebrates to pump hemolymph into a coelom where it diffuses back to the heart between the cells.

The closed system necessitated the development of thin-walled capillaries to allow exchange between the blood and lymph. This closed system is used by fish and all its more sophisticated descendants, the amphibians, reptiles, birds and mammals.

This was the completion stage of the basic animal blueprint for that followed.

---

[1] https://cbs.umn.edu/cgc/what-c-elegans

INSECT EDEN

The next great step in animal evolution was the Logos-directed duplication of the basic pattern followed by different developments of the duplicates. The first multi-segmented animal looked like a set of worms all joined together. Each segment was essentially complete with only the external skin, the gut and blood vessels connecting them. At this time appeared a genetic innovation still active in all sophisticated animals: the regulatory homeotic genes, all sharing a highly-conserved 180-base sequence called the homeobox. These homeotic genes control the duplicated segmentation:

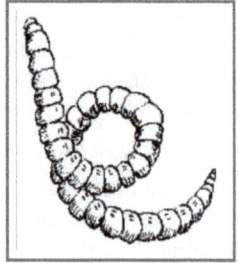

> "{This was} a major mechanism of evolutionary diversification, perhaps the most important one in the history of life. It initiated an extraordinary combinational [process] involving complete, originally viable modules that could be mutated, fused, deleted and otherwise reshuffled, all by the magic stroke of a single or sparse genetic modification..."[1]

These simple segmented worms are extant as the annelid worms, the most familiar example being the common earthworm which has adapted to dry land, with a mouth in its anterior segment and an anus in its posterior segment.

The differentiation of the segments developed over time leading to whole classes of invertebrate animals that are predominant on Earth to this day—shrimps, lobsters, crabs et al. the ocean, and on land, insects, spiders, scorpions et al.

The fruit fly that seems to appear out of nowhere when ripe fruit is left lying around is the favored subject for geneticists. It was here that the homeotic genes were discovered.

> "Scientists discovered homeotic genes by studying strange transformations in fruit flies, including flies that had feet in place of mouth parts, extra pairs of wings, or two pairs of balance organs ... instead of wings. Some even had legs growing out of their heads in place of antennae!"[2]

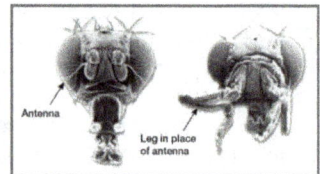

While it is difficult, nowadays, to discern the segmentation stage in human development, the clearest remaining representation of this aspect of the Logos in human development is our spinal column that, in the embryo,

---

[1] Christian de Dove, *Vital Dust,* BasicBooks 1995, p.197

[2] http://learn.genetics.utah.edu/content/basics/hoxgenes/

starts as the cartilaginous notochord that is calcified into the set of segmented vertebra.

The final change that opened up the possibilities of modern sophisticated animals was so strange that it can only be ascribed to Logos forward-looking to future development. The mouth end and the anus end switched roles.

The insects and all the others remained as always, and are called the protostomes—the mouth-first organisms. The arthropods had the easiest transition to land already covered with an impervious chitin carapace. The only crucial step was to replace the external feathery gills with an internal set of ramifying tubes reaching bringing air to all parts of the body. This eventually led to the thousands of different insects and their relatives that rapidly followed plants onto land. As sucking plant liquids is not that different from sucking animal blood, mosquitoes and their ilk developed and have been a nuisance ever since.

FISH EDEN

One lineage, however, switched the roles of the holes—with no discernible role in fitness as required by Darwin's random variation—so that the ancient anterior first-hole switched from being the mouth to the posterior anus—while the newcomer second hole, presumably not so set in its ways, switched from being the anus to become the anterior mouth. This homeotic Logos-guided 'flipped' deuterostome lineage became the ancestors to all vertebrate fish, amphibians, reptiles, birds, mammals and humans.

It is possible that, without this fateful switch, there would be no fish, no amphibians, no reptiles, no birds, no mammals, no humans.[1]

A consequence of this flip, still in evidence in simple acorn worms that have changed little over time, was that the anterior end became responsible for food and oxygen uptake. The front end developed into a set of gill slits and water taken in through the mouth was forced out through the slits that absorbed water and trapped food particles. These gill slits are clearly seen to this day in the embryonic development of all vertebrate embryos.

Another step was an infolding of the ectoderm along the dorsal skin which became a hollow tube running along the back, while below it a cartilaginous rod developed—the notochord—that is the signal characteristic of all chordates. This appears, if transiently, in

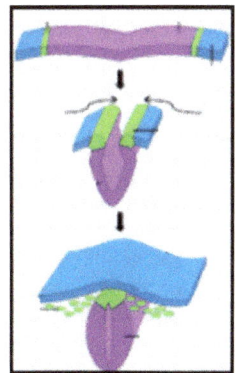

---

[1] Christian de Duve, 1995, *Vital Dust: Life as a cosmic imperative*, Basic books p. 199

all embryonic and some adult chordate animals. An extant form of this developmental stage are the lancelets, fish-like marine chordates with a global distribution in shallow temperate water, usually found half-buried in sand.

The next step was to surround both the neural tube and the notochord with segmented cartilage structures that protected the proto-spinal cord while remaining flexible.

"The first vertebrates had cartilaginous bones and resembled worms more than fishes, having no jaw and only rudimentary fins. According to the fossil record, some were bizarre, ferocious-looking animals covered with armored plates. Their closest present-day descendants are the lampreys and hagfishes, which are very different from the remote relatives but share some primitive features with them."[1]

The next major advance learnt from the Logos was the formation of the hinged jaw by reforming the cartilage supporting the foremost gill slits. With the development of fins supported by cartilage and the muscles to move them, the cartilaginous fish, including the sharks, rays, and sawfish, a class distinguished by having a skeleton of cartilage rather than bone.

The final and completion stage for fish was seeding the cartilage structures with calcium salts converting the cartilage into bone. Nowadays, about 90% of the world's fish species are of the bony fish class.

Up to this point in history, the only organisms to thrive on dry land were the plants, fungi and insects. Now it was the turn of fish to make the transition.

Land Animal Edens

The time-honored phrase, *fish out of water*, is an idiom used to refer to a person who is in unfamiliar, and often very uncomfortable, in new surroundings. This suggests the many challenges faced by the bony fish—gravity not supporting the cartilaginous varieties—in surviving in air and meagre water.

Sensibly, they did this in gradual stages, the first being the amphibious vertebrates, that spent intervals in both air and water, and have remained successful to this day. While the details of the transitional forms from fish to frog are still murky, it is widely accepted to have involved the lobe-fin fish:

"The most important features of lobe-finned fish is the lobe in their fins. Unlike other fish, Lobe-finned fish have a central appendage in their fins containing many bones and muscles. The fins are very flexible and potentially useful for supporting the body on land, as in lungfish and tetrapods (vertebrates with four limbs). Tetrapods are thought to have evolved from primitive lobe-finned fish."[2]

---

[1] Christian de Dove, *Vital Dust*, BasicBooks 1995, p.201

[2] http://www.mesa.edu.au/fish/fish04.asp

While prominent in the fossil record, it was thought for many years that relatively-direct ancestors had gone extinct until the fortuitous discovery of a living example, the coelacanth:

"The coelacanth was thought to have become extinct 65 million years ago until its capture in 1938 by a South African museum curator on a local fishing trawler…. The most striking feature of this "living fossil" is its paired lobe fins that extend away from its body like legs and move in an alternating pattern, like a trotting horse. Other unique characteristics include a hinged joint in the skull which allows the fish to widen its mouth for large prey"[1]

The fossil record reveals much about the transition:

"There is a sequence of fossils which occupy the transition from fish to amphibian. 378 mya… These are lobe-finned fish. … The skull bones of these fish are bone for bone equivalents to the skull bones of the earliest tetrapods…. This fish also had lungs and nostrils … but also had gills. These things really looked like tetrapods until you see the fins."[2]

AMPHIBIAN EDEN

Stranded, or visiting, the land from water, these fish used their swim bladder to survive, a characteristic leading towards an amphibian toleration of air, and extant as the lungfish:

"There are a number of fishes that, in addition to or in place of gill breathing, have developed special organs through which they can breathe atmospheric air at the water surface. This occurs almost exclusively in freshwater fishes. In lungfishes these organs are, both in function and in structure, primitive lungs like those of amphibians. The name lungfish is thus well applied: these fishes have lungs that are derived from the swim bladder (an organ used for buoyancy in most bony fishes), which is connected to the alimentary tract. The inner surfaces of these air-breathing organs are covered with a great number of honeycomb-like cavities supplied with fine blood vessels. As in terrestrial higher vertebrates, gas exchange takes place in tiny air vesicles. Also as in terrestrial vertebrates, there is a separate pulmonary circulation."[3]

The great reward that encouraged this development was the abundant food plants on land as the Carboniferous plants flourished, only nibbled at by the contemporaneous insects.

Many species of amphibians were terminated in the great Permian Extinction that followed but many survived. Some retained their tails—such as newts and salamanders—while others went through a metamorphosis from fish-like larvae to tailless-forms such as the extant frogs and toads where the tail is absorbed and legs sprout, a change governed by thyroxine, an iodine-

---

[1] http://chem.tufts.edu/science/evolution/fish-amphibian-transition.htm

[2] *ibid*

[3] https://www.britannica.com/animal/lungfish

containing hormone, that is crucial in vertebrate development to this day. The lack of which causes problems in human development:

"Thyroid hormones are critical for development of the fetal and neonatal brain, as well as for many other aspects of pregnancy and fetal growth. Hypothyroidism in either the mother or fetus frequently results in fetal disease; in humans, this includes a high incidence of mental retardation."[1]

Another innovation, was the widespread control of programmed cell death—called apoptosis—during development, such as the buds at the ends of the lobes which are carved by cell death into the familiar fingers and toes.

"Apoptosis is a form of programmed cell death, or 'cellular suicide.' It is different from necrosis, in which cells die due to injury. Apoptosis is an orderly process in which the cell's contents are packaged into small packets of membrane for 'garbage collection' by immune cells. Apoptosis removes cells during development, eliminates potentially cancerous and virus-infected cells, and maintains balance in the body."[2]

REPTILE EDEN

The factor that kept amphibians tied to wet-lands was that their reproduction required expanses of water, a distinct disadvantage to reproduction during the Permian drought. The Logos provided a solution that was rapidly adopted: the zygote surrounded, protected and nourished by the egg:

As happened many times, life rallied; evolution responded to ecological challenges by appropriate adaptions. It even turned disaster into success driven, by the great Permian crises to accomplish one of the most decisive advances. While seed plants took over the cold, dry swamps left barren by the decimation of sporulating plants, some obscure amphibian suddenly soared into prominence by developing the animal equivalent of the seed: the fluid-filled egg[3]

The normal amphibian mode of reproduction was to release multitudes of zygotes into water where at least a few would survive to form the next generation. Guided by the Logo, a female started a new lineage. In this lineage, the zygote was enclosed:

"In a fluid-filled sac, the amnion, within which the embryo could pursue it's normal aquatic development.... a *milieu exterieur* to shelter the developing embryo. A hard, porous shell protected this substitute marine incubator, while a highly vascularized membrane, the allantois, produced by the embryo, and lining the inner face of the shell, served in gas exchange and waste disposal. Another sac, filled with a richly nutritious yolk, provided the embryo with the necessary foodstuffs. Thus, the complete development of the organism up to the stage where it could survive on land took place within

---

[1] http://www.vivo.colostate.edu/hbooks/pathphys/endocrine/thyroid/thyroid_preg.html

[2] https://www.khanacademy.org/science/biology/developmental-biology/apoptosis-in-development/a/apoptosis

[3] Christian de Dove, Vital Dust, BasicBooks 1995, p.207

the protective, well-stocked and appropriately renewed environment of the amniotic fluid True terrestrial reproduction was initiated. The first reptile was born"[1]

This advance proved its value during the Permian crises when reptiles advanced and radiated in a variety of forms, some of which abandoned legs, including the extant lizards, snakes and turtles—that live in the ocean but in a anti-amphibian manner return to land for breeding—but by far their greatest impact on history (and young human males) was as the sensational array of dinosaurs. These are so well-known that it is not necessary to discuss them here.

BIRD EDEN

Some dinosaurs took to the sky—the pterosaurs—with wingspans of 30 feet or so. It was the birds, however, that are the extant descendants of the dinosaurs:

Landed on the world some 150 million years ago, as revealed by the famed *Archaeopteryx*, a fossil discovered in 1864 in a schist quarry in Bavaria. This weird animal would have passed for a small dinosaur by any test if it were not for the imprint of feathers miraculously preserved in the soft stone. Feathers, indeed, turned a reptile into a bird.[2]

This quite unexpected connection was relatively quick as it was guided by the Logos with birds as the goal. This rapid change is a puzzle to the random chance-and-accident considered dogma to many scientists:

"Evolutionary novelties were appearing on the bird stem lineage at a faster rate than across the rest of the tree. Many were major innovations such as complex feathers, bigger brains, wings and wishbones. Stem-birds were out-evolving their contemporaries by changing approximately four times as fast. This continual and often rapid shrinking was probably directly related to the accelerated evolution of anatomical novelties. Reduced body size, for instance, allowed bird-stem dinosaurs to explore new postures (bird-like walking where the thigh bone is held horizontal) and habitats (such as arboreal and, later, aerial habitats). This in turn would have created pressure to evolve radical new adaptations such as reshaping fluffy feathers into wings."[3]

The land-animal history, from the first amphibian appearance some 400 mya to the great dinosaur extinction 65 mya, is well-documented. In the

---

[1] Christian de Dove, *Vital Dust*, BasicBooks 1995, p.207

[2] ibid, p. 210

[3] http://www.iflscience.com/plants-and-animals/how-small-birds-evolved-giant-meat-eating-dinosaurs/

fossil record, the dinosaurs disappeared along with the heretofore plentiful ammonite mollusks and the temporary replacement of flowering plants by ferns.

This holocaust is thought to have been caused by an 6-mile-diameter asteroid hitting the Yucatan Peninsular in Mexico with an energy equivalent to 100 million megatons. The following 'nuclear winter' caused by the large amounts of sooty smoke ejected globally into the upper troposphere and lower stratosphere, reflecting sunlight and cooling the planet.

Mammalian eden

In what philosophers might call the victory of maternal tender-loving-care over might-is-right, a branch of the heretofore cold-blooded dinosaurs learned from the Logos the ability to thrive in cold climates by keeping their internal temperature at an optimal ~100°F and the faculty for gaining the extra food needed to fuel this warmth by carnivorous hunting, a thick pelt of fur and females who kept the eggs warm, sheltering the newborns. Nourishing the young by secreting fatty liquid from their chest, this eventually developed into the mammary glands and the advent of maternal care by the mammals.

They were actually around for ~200 million years during the ascendancy of the dinosaurs, rat-like and rarely bigger than a rabbit. One lineage learnt how to hatch the eggs inside the female body. The marsupials—such as the kangaroo—delivered the very immature young and nourished them in an external pouch about the mammary glands. The final reproductive development was the placenta which allowed internal development so that the young were often essentially functional at birth—such as a foal that can walk just 15 minutes after birth—or need a period of parental care that—as in the case of humans—might last a decade.

"They acquired certain traits that would characterize mammals ever afterward: limbs positioned under the body, an enlarged brain, a more complex physiology, milk-producing glands, and a diverse array of teeth -- incisors, canines, premolars, and molars. Already present were the ancestors of the three major mammalian groups that exist today— monotremes (platypus and spiny anteater), which lay eggs externally; marsupials (kangaroos, opossums), which carry their young in a pouch; and placental mammals (humans, cows, horses), which retain the fetus internally during long gestation period. In the early Cenozoic era, after the dinosaurs became extinct, the number and diversity of mammals exploded. In just 10 million years—a brief flash of time by geologic standards —about 130 genera (groups of related species) had evolved, encompassing some 4,000 species.[1]

---

[1] https://www.pbs.org/wgbh/evolution/library/03/1/l_031_01.html

Human eden

Genetics reveals many things, one being the relationship of various lineages. This leads to a controversial, if provable, fact of human lineage;

The chimpanzees are the closest. Just 6 million years ago, a single female ape had two daughters. One became the ancestor of all chimpanzees, the other is our own [great ×] grandmother.[1]

There was a long history leading up to origin of modern humans. One of the earliest fossil of this preparation period is that of famous Lucy:

"*Australopithecus afarensis* is one of the longest-lived and best-known early human species—paleoanthropologists have uncovered remains from more than 300 individuals! Found between 3.85 and 2.95 million years ago in Eastern Africa (Ethiopia, Kenya, Tanzania), this species survived for more than 900,000 years, which is over four times as long as our own species has been around. It is best known from the sites of Hadar, Ethiopia …'Lucy'… and the 'First Family'…; and Laetoli fossils of this species plus the oldest documented bipedal footprint trails"[2]

The dating of the fossil records found in Africa reveals the timing—in millions of years ago (mya)—of the steps towards our fully-functioning upright posture that the pre-human lineage learnt from the Logos:
Ability to walk upright (6 mya); Strong knees (4.1 mya); Curved spine (2.5 mya); Hip support (2 mya); Fully bipedal (1.9 mya).[3]

About the time a lineage became fully bipedal, history entered the Old Stone Age, the Paleolithic Period: "[Pre] humans in East Africa used hammerstones to strike stone cores and produce sharp flakes. For more than 2 million years, early humans used these tools to cut, pound, crush, and access new foods—including meat from large animals."[4]

It was thought that pre-humans only mastered fire only towards the end of the Paleolithic but discoveries in 2012 suggest it happened about halfway into this period:

"Now, however, an international team of archaeologists has unearthed what appear to be traces of campfires that flickered 1 mya. Consisting of charred animal bones and ashed plant remains, the evidence hails from South Africa's Wonderwerk Cave, a site of [pre-human] habitation for 2 million years."[5]

---

[1] Harari, Yuval Noah. *Sapiens* (p. 5). Harper.

[2] http://humanorigins.si.edu/evidence/human-fossils/species/australopithecus-afarensis

[3] http://humanorigins.si.edu/human-characteristics/walking-upright

[4] http://humanorigins.si.edu/human-characteristics/tools-food

[5] http://www.history.com/news/human-ancestors-tamed-fire-earlier-than-thought

Stone tools and fire were the main advances during the 2 million years of the Paleolithic, the Old Stone Age.

[Humans] replaced all the hominid populations without merging with them. If that is the case, the lineages of all contemporary humans can be traced back, exclusively, to East Africa, 70,000 years ago.[1]

This 3 million year stasis of the Paleolithic ended and we entered a time of rapid development: the New Stone Age (Neolithic), advent of agriculture, the first cities, pottery, the Bronze Age, writing, the Iron Age, etc. This period of rapid change was initiated by the Origin of Humans, the topic of the next chapter.

While the details of this evolutionary leap are still under investigation, one thing is clear: On the path of ape to human, a major chromosomal rearrangement occurred.

Recently, with the determination of the complete sequence of the human genome, it has become possible to look at the precise location where this proposed chromosomal fusion must have happened. The sequence at that location—along the long arm of chromosome 2—is truly remarkable. Without getting into the technical details, let me just say that special sequences occur at the tips of all primate chromosomes. Those sequences generally do not occur elsewhere. But they are found right where evolution would have predicted, in the middle of our fused second chromosome. The fusion that occurred as we evolved from the apes has left its DNA imprint here. It is very difficult to understand this observation without postulating a common ancestor.[2]

_____

[1] Harari, Yuval Noah. *Sapiens* (p. 15). Harper.

[2] Collins, Francis S.. *The Language of God: A Scientist Presents Evidence for Belief* (p. 138). Free Press.

8 • THE ORIGIN OF MAN

This is probably the most contentious chapter because, while folks do not usually get emotional about Big-Bangs, quarks or wavefunctions, they do when the question of human origins is explored. As discussed in Chapter 1, the relation between God and Creation is that God does not control what happens in the physical world because, by design, He intended True-love Man to be Lord of the Physical World. Before the advent of Man, God rules the physical world indirectly through the Logos; it is in the 'indirect dominion.'

When True Man emerges and rules the physical world with true love, the Logos continues to run the 'housekeeping' aspects of Creation such as thermonuclear fusion in the sun and the weather. The closest concept that current science has to the Logos is 'natural law.'

Unification Thought emphasizes that there are two distinct stages in God's Creation of the universe as illustrated here.[1]

First, the Logos was created starting from top to bottom, from Man via animals and bacteria, all the way down to atoms and spacetime.

Second, the Logos was put into action with the creation of spacetime in the Big-Bang, and the Logos was then expressed over time, step by step via atoms, bacteria and animals until it reached its complete expression in the birth of human beings.

The relation between these two stages is akin to the much simpler relation between creating the idea of *Ode to Joy* in the mind of Beethoven, and the energy and effort it took to write the score and assemble an orchestra to perform it.

Logos and Origin

As discussed earlier, the origins of inanimate systems and the first of an animate systems are the same: The Logos is expressed through develop-

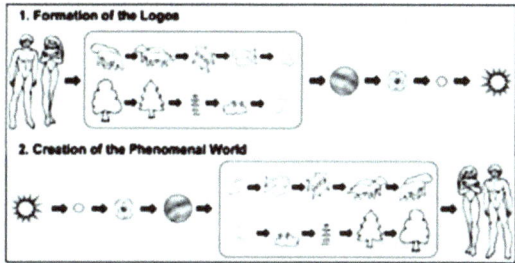

---

[1] Akifumi Otani, *Beyond Darwinism: Towards Unification Science*, 2010, Unification Thought Institute, Tokyo, Japan. Available in PDF form (large file) at:
http://www.utitokyo.sakura.ne.jp/uti-index-gaiyou01-siryou01-2011-e-01

ing and merging the wavefunctions of interacting subsystems into the wave-function of the emergent new system with an attendant set of new emergent properties.

In living systems, the novel analog form of the interacting subsystems is captured as digital information stored in the DNA as discussed earlier. The origin of the second, third, etc., of the living system is qualitatively different in that the Logos is not directly involved; it is the digital information that directs the assembly. The origin of this lineage, however, is quantitatively the same as the first origin as the same analog form and emergent properties result.

This direct involvement of the Logos is what Unification Thought calls a "special creation" as distinct from regular repro-duction.

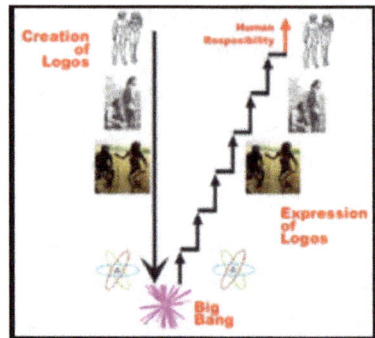

> And even though the stages of ape-men and early men were traversed before reaching the stages of homo sapiens, nevertheless there must have been a great leap when human beings (i.e., Adam and Eve) were created.[1]

Completion of the Logos

As discussed earlier, the Logos is sequentially expressed in living systems over time, passing through what can be colloquially called the Age of Bacteria, the Age of Protists, the Age of Worms, The Age of Fishes, the Age of Reptiles, the Age of Mammals, and the Rise of the Primates.

The focus here is on he final stage, when the Logos reaches its complete expression in the birth of the first human pair, Adam and Eve. This is God's 95% responsibility. Unification Thought states that the final stage of maturing to express the love of God, the final 5%, is a creative act of human will, the fulfillment of human responsibility as they develop.

The exact sequence of history is still a matter of debate so, for convenience, we will equate the pre-human stage with the Neanderthals. There was, as usual, preparations for this final stage of evolution:

> Some [hominids] may have made occasional use of fire as early as 800,000 years ago. By about 300,000 years ago, Homo erectus, Neanderthals and the forefathers of Homo sapiens were using fire on a daily basis. Humans

---

[1] Sang Hun Lee, supervisor, *From Evolution Theory to a New Creation Theory*
http://www.tparents.org/Library/Unification/Books/EvolTheo/EvolTheo-03.htm

now had a dependable source of light and warmth, and a deadly weapon against prowling lions.[1]

Materialism has that the transition from Neanderthal to human was a gradual event over an extended period of time. The science of Godism has that there was a specific speciation event in which epigenetic information, gleaned from the Logos via the environment, directed a reconfiguration of the Neanderthal genetic material into the human configuration. A new set of qualities was inherited from the Logos, including the knitting together of the mind into the "I Am" sense of self. This ultimate emergent property also imposed upon the spirit self an eternal existence, that is not an attribute of lesser organisms.

The speciation event always involves a male and female, and the result was the birth of Adam and Eve, as the first humans are traditionally called. Classical science states that the transition from Neanderthal to Human was caused by random variation and survival of the fittest. Unification Thought states that speciation occurs by a specific mechanism directed by accumulated epigenetic ancestral wisdom inherited from Logos passed down a lineage to the first humans.

Abilities that for the Neanderthals were learnt with great effort became hard-wired into humans. The mental abilities of a mature Neanderthal were comparable to those of a six-year-old human, and they could develop no further. In yesterday's psychology they would have been classified as Imbecile (above an idiot and a below a Moron). There is evidence that various races of Neanderthals, specialized for various environments, came together and 'pooled their wisdom' in the lineage leading up to humans.[2]

Neanderthals probably communicated with a simple pidgin of nouns and verbs[3], had mastered fire, and were successful gatherers and hunters wielding crudely chipped lumps of flint rock.

The time and place in which the first humans were born is traditionally called the Garden of Eden and science has a rough idea of when it occurred; the transition from the Paleolithic age (Old Stone Age) to the Neolithic age (New Stone Age).

---

[1] Harari, Yuval Noah. *Sapiens* (p. 12). Harper.

[2] https://www.nature.com/articles/507303a.

[3] https://www.sapiens.org/column/field-trips/did-neanderthals-speak/

Old and New

We will equate the period in which the neanderthals were the most sophisticated of the primates with the Paleolithic age, which lasted about 2,500,000 years. The cultural changes over this vast stretch time were very slow and incremental, with the controlled use of fire appearing just 400,000 years ago.[1] The sophistication of the stone implements, for example, hardly changed over millions of years. Stasis and equilibrium were the rule.

Archeological evidence exists that shows that they buried their dead and eventually had mastery of fire for cooking. They were communal (clan and tribal level) and probably had a pre-language (pidgin) of simple nouns and verbs. They fashioned simple stone and bone tools, and were successful hunters and could fend of predators such as the great cats. The fossilized Laetoli footprints left in 3,000,000 year-old volcanic ash, footprints of a pre-Neanderthal male, a female and a child[2], suggest that pair-bonding reproduction was already established at this early stage.

This multimillion-year stasis was punctuated by the emergence of Man and the start of the neolithic age ~100,000 years before the present (YBP) and was in full swing ~50,000 YBP. The stone and bone shaping of tools was much more sophisticated and decorated. They had a true language of syntax and grammar.[3] The hunter-gatherer stage developed into that of agriculture and the domestication of animals >20,000 YBP and, most distressingly, the earliest evidence of a multi-person battle is 14,000 YBP.[4]

Habitations beyond caves were developed. The discovery of how to smelt copper from its ores and how to create its alloy, bronze, marked the end of the neolithic and the start of the Bronze Age ~15,000 YBP. Writing was developed soon after. Unlike the million-year stasis of the Old Stone Age, innovative change over thousands of years was the rule in the New.

Assuming a new set of emergent properties happened; where did it happen? The location of the Origin of Man, the "Garden of Eden," has been roughly established by three lines of evidence that are all in essential agreement. These are the study of the female lineage using mitochondria, the study of the male lineage using the Y-chromosome, and the spread of languages around the world.

---

[1] http://en.wikipedia.org/wiki/Control_of_fire_by_early_humans

[2] http://www.pbs.org/wgbh/evolution/library/07/1/l_071_03.html

[3] https://gawron.sdsu.edu/fundamentals/course_core/lectures/historical/historical.htm

[4] http://en.wikipedia.org/wiki/Cemetery_117

Lineage & Linguistics

The history of the human female lineage is tracked by tracing the spread of genetic markers on the mitochondrial chromosome which is passed down the female lineage from mother to child. The mitochondria are not passed on by males. If a mitochondrion does make it from the sperm into the egg, it is immediately surrounded and destroyed.[1]

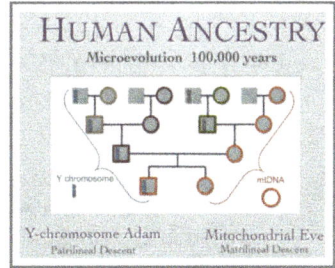

The pattern of human migration that emerged from these studies was that the female lineage started off in East Africa, then humans spread south into Africa and north to the rest of the world.[2] This original female is called "Mitochondrial Eve" in the literature.

The history of the male lineage is tracked by tracing the spread of genetic markers on the Y chromosome which is passed solely down the male line from father to son, and is not passed on to females. The pattern of human migration emerged was that it started off in East Africa, then the humans spread south into Africa and north to the rest of the world.[3] This original male is called the "Y-chromosome Adam."

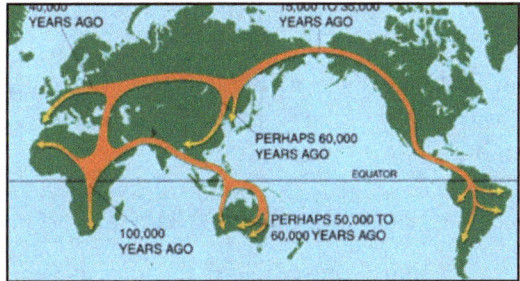

The current view of gradual speciation has the mitochondrial Eve and the Y-chromosome Adam as members of a small breeding population. The unified view of a directed, specific mechanism for speciation has that this population was as small as two.

For a while, it looked as if this genetic Adam and Eve existed far apart in time, but advances in genetic chronology have now placed them close together in time: "Now, two major studies of modern humans' Y chromosomes suggest that 'Y-chromosome Adam' and 'mitochondrial Eve' may have lived around the same time after all... 180,000 to 200,000 years ago."[4]

---

[1] http://www.nature.com/nature/journal/v402/n6760/full/402371a0.html

[2] http://marioarland.edublogs.org/2008/06/26/recent-single-origin-hypothesis/

[3] http://t2.gstatic.com/images?q=tbn:ANd9GcQvRNF8yIFl_XD5ugG9Ui5Smza--sXw85s-Y8ZFdhbm-vnXVSqprwQ

[4] https://www.nature.com/news/genetic-adam-and-eve-did-not-live-too-far-apart-in-time-1.13478

Language has changed over time and migration, and seems to have a single origin:

> Many linguists believe all human languages derived from a single tongue spoken in East Africa around 50,000 years ago.[1]

The study of how language has changed over time as humans migrated from East Africa shows a similar pattern to the genetic studies:

> "A new linguistic analysis attempts to rewrite the story of Babel by borrowing from the methods of genetic analysis—and finds that modern language originated in sub-Saharan Africa and spread across the world with migrating human populations."[2]

All three lines of investigation suggest that the eden into which the first humans were born was in East Africa less than 100,000 years ago. Genetic analysis of the genes for skin color indicate that the first humans were black, and that the yellow and white pigmentation arose much later as human migration progressed.[3]

There is evidence that different races of Neanderthal commingled on the route to Human, including races adapted to water as well as other races adapted to forrest and savanna.[4] The advantages that occur in such 'miscegenation' is known as 'hybrid vigor' in practical genetics.[5]

On purely esthetic grounds, I like to think that the Neanderthal tribes gathered, interbred and gave rise to humans in East Africa's most dramatic and bountiful landscape between Mt. Kilimanjaro and the Great Lakes.

Completed Creation

Unificationism rejects the idea that the first humans were created fully formed, and accepts that they emerged much as babies emerge today. Basic physiology was then as it is now, so we can infer much about their situation.

Putting this altogether, we conclude that the final stage of Creation, the expression of the Logos, was completed with the birth of the first two humans and their natural development to about 6 years old and the Nean-

---

[1] https://www.livescience.com/16541-original-human-language-yoda-sounded.html

[2] http://aminotes.tumblr.com/post/4633090702/evolution-of-language-tested-with-genetic

[3] http://barclay1720.tripod.com/hist/origin/outafrica.htm

[4] http://en.wikipedia.org/wiki/Aquatic_ape_hypothesis

[5] http://www.thefreedictionary.com/hybrid+vigor

derthal level of personality. Their development from then on to maturity with the ability to love as God loves was their own creative responsibility.

We can assume that Adam and Eve were born into the midst of a flourishing and supportive tribe of Neanderthals. They had a biological father and a biological mother,[1] and furthermore, as these parents were at the highest level of Neanderthal development, we can assume they had a high status in the tribe; perhaps even its leaders. As is normal for all human beings, the first two humans emerged as helpless newborns and they were fed at their Neanderthal-mother's breast.

> Now, let me talk about the history regarding Adam's birth. Did Adam have a belly button or not? You must know it. Without a belly button, where was he born from? Adam had a navel cord, and he had a mother.
>
> Sun Myung Moon *The Blessed Family*, 1999 winter issue

As they grew, they were protected from predators and provided with food. They played with Neanderthal children, and were taught the ways of the tribe. They were similar to their peers in their capacities until they were past the age of six. Then they left their peers way behind them.

In particular, human children have the innate capacity, emergent from the Logos, to create a true language out of a pidgin, a language with grammar and syntax.[2] So Adam and Eve would have created their own language, a level of sophisticated communication unavailable to the Neanderthals, the 'ur-language' as it might be called.

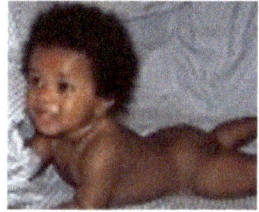

The first humans were two golden children maturing in a world of kind and supportive imbeciles. While the Neanderthals were simpletons by human standards, they were perfectly functioning as they were intended, and they had their own wisdom of how to live successfully.

Role of angels

Many religions, including Unificationism, include angels, as well as God and humans, in the parables about human origins. They are usually considered to be God's helpers and assistants.

What kind of assistance would be helpful to the Creator God. Surely not in mastering calculus, or figuring out the exact balance between the electromagnetic and gravitational forces so that the Sun is possible. No, in all

---

[1] Sun Myung Moon, The Blessed Family Quarterly, 1999 winter, 28.

[2] Steven Pinker *The Language Instinct*, 1994.

that has been said about Adam & Eve, by far the most useful thing angels could do for God are the things He cannot do Himself, such as:

1. Give birth to Adam & Eve
2. Suckle them with mother's milk
3. Protect them from lions, tigers and bears, etc.
4. Wean them, and then feed them nourishing food
5. Teach them toilet discipline, etc.
6. Tell them to stay home, not go off exploring

We reach the conclusion that angels are symbols for the Neanderthals that did all of this for Adam & Eve. The concept of angels existing with God from the very beginning is in contradiction with the two-stage creation, and the stepwise expression from simple to complex. Angels, being just less than human, could not have appeared first, and Unification Thought gives angels no role to play alongside the Logos.

Our mental picture of angels is not the Sistine Chapel but rather a Neanderthal diorama.

Only one other entity besides God, Adam and Eve is mentioned in the Bible; Lucifer, an archangel, a leader of the angels in a position to influence Adam and Eve.

From what has been gone before, we can speculate that Lucifer is the male Neanderthal who's mate gave birth to the first humans, and that Lucifer is a symbol for the biological father of

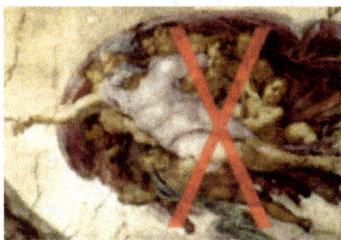

Adam and Eve. It was he who raised the first humans, and gave them the Commandment that was to protect them.

In the Bible stories we see the important, also-starring role of women in the Providence • Noah's wife (not supportive) • Abraham's wives (Sarah (ancestor of the Jews/Judaism) expelled Hagar (ancestor of Arabs/Islam) • Rebecca (good, helped son Jacob fool husband) • Elizabeth (mother of John-the Baptist) expelled pregnant Mary (mother of Jesus) creating a deadly situation culminating in the Crucifixion.

For all these roles in **recreation**, Eve is the only female role mentioned in original **creation**. Where is the female figure who did all the hard work of giving birth to the two babies? Who did all the motherly work of raising them? Whose breasts suckled them for many months? Who kept an eye on her husband, their biological father?

Lucifer, the kind, imbecile father figure was intended in God's plan to fall totally in love with his beautiful children, to fall in love with Adam and Eve. The love that developed in the male parental Neanderthal would be balanced and healthy, just as God intended. Adam and Eve were to grow and develop in this environment to maturity and completion of the purpose of creation. First, we will discuss what God intended for His first children; then we will discuss what could possibly have gone so wrong as to infect all of humanity since with a broken capacity for love.

Intended Ideal

Adam and Eve were to grow to perfection, to love as God loves, with unconditional parental love for all. They were to grow through the stages of love capacity, using their free creativity until their love was mature.

To guide them along this path without mishap, the Bible parable states that God gave a commandment, and their responsibility was simple; to obey it. The Commandment could have been as simple as: *Stay at home*, and it would quite natural for this admonition to come from the parental Neanderthals.

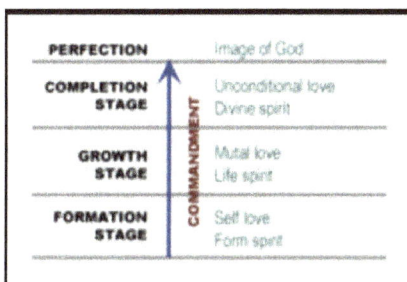

PERFECTION	Image of God
COMPLETION STAGE	Unconditional love / Divine spirit
GROWTH STAGE	Mutual love / Life spirit
FORMATION STAGE	Self love / Form spirit

There is support for this concept in the adage: *Two's company, three's a crowd*, and why, in this age of sex abuse, priests and teachers are advised to always have another person present. It is the difference between a private and a public relationship.

Indirect support for this simple Commandment is given in an early systemization of Unificationism when discussing how Lucifer's love-to-lust occurred:

> If Adam had watched over Eve more closely and spent more time with her, this would not have happened.[1]

In this balanced situation Lucifer would be dazzled by the brilliance of his children and nothing untoward could occur, and the purpose of Creation would be completed with the uniting of Adam and Eve in True Love.

To put it simply, the Commandment was to keep them out of trouble, to keep them safe. In this situation, their character and ability to love, their spirit, would have naturally grown from self love, through mutual love to unconditional love, God's love. From self-centered spirit, through a Form

---

[1] http://www.tparents.org/Library/Unification/Books/DpStudy1.pdf p.83

and Life spirit, to a Divine spirit. They would live out their days on earth, raise a true family of children and grandchildren, and then, discarding their physical bodies, pass into the spirit world where they would spend the rest of eternity with God in the realm of True Love, the kingdom of heaven. The nature of the Spirit realm, its substantial aspect and relationship to recent advances in science has been discussed earlier.

With Adam and Eve raising their family in true love, there was no more necessity of the Commandment as Adam and Eve were quite capable of keeping their children out of trouble.

The humans race would multiply and develop true love culture as the natural leaders of the Neanderthals, who would greatly proper under their loving and creative care. This is the dominion of true love humans over Creation, and is the fulfillment of the Third Blessing.

One rather surprising implication of this view is that the first animal to domesticated would be the Neanderthal, not the dog. The Neanderthals would be the natural servant class, while the true love humans would be the natural aristocrats. With plenty of leisure time, humans would rapidly develop agriculture, science and an abundance of art.

The image of a natural, hereditary aristocrat and servant class tends to be associated with the endless examples of unnatural slavery in human history. The images of White master beating Black slave, or Japanese overlord and Korean underdog are two not-so-distant examples of attempts to enslave human beings. The true love relationship is quite different to these shudder-inducing examples from fallen history.

The relationship of Human lord and domesticated Neanderthal is the ideal portrayed on American TV by the relationship between Joe and his dog Lassie, and the Lone Ranger and his horse Silver. The image is not master-beating-slave but rather, "Dogs who talk, do dishes and can be toilet-trained."

A 'slave rebellion' would be as unthinkable as is a pedigree-dog revolt. The true love culture would fill the earth, and then spread out to all the galaxies.

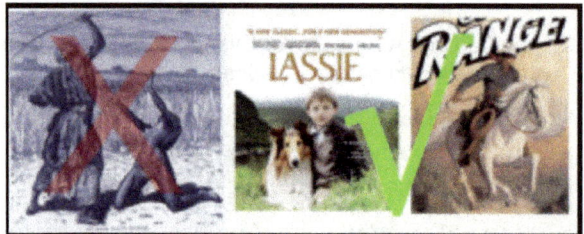

What Went Wrong?

This loving, leisurely civilization never emerged. Instead a brutish culture developed where there was no true love and all men were out for themselves. Family dysfunction was the norm—see King Lear or all of classical Greek plays for examples. It was a culture into which "Thou shalt honor your father and mother" had to be injected as a divine revelation. Life was nasty, brutish and short; and the spirit of those times has been chillingly recreated in a movie about slaves and human sacrifice.[1]

Unificationism states that there were two stages in the Fall of Man:

1.   Adam did not obey the Commandment. In fact the Bible relates that he went off naming all the animals. He did not stay home, and the father-Neanderthal was left with only Eve to fall in love with.

2.   Lucifer was twisted into Satan by the misdirected power of love, and in this state had a sexual relationship with Eve.

Because Adam disobeyed, the Neanderthal headman was left alone with an enchanting, beautiful young woman, Eve. This was an unbalanced situation and Lucifer's love for Eve became twisted into lust. His mind was filled with things that had nothing to do with the Logos which he had resonated with before.

Now we have to ask the disquieting question: what kind of sexual relations did this parental, grown male have with the young woman. What kind of sexual relation could have so brutalized Eve that she brutalized her children and her children's children down to the present day.

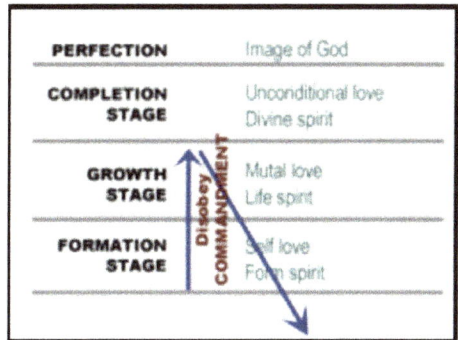

PERFECTION	Image of God
COMPLETION STAGE	Unconditional love / Divine spirit
GROWTH STAGE	Mutal love / Life spirit
FORMATION STAGE	Self love / Form spirit

For an answer to this we turn to what is known from human psychology about the ubiquitous and deeply-disturbing prevalence and consequences of parental figures raping their daughters.

Child Rape

It is only in these supposedly best-of-times that the prevalence of father-daughter rape has become well known; we can assume that it was even more prevalent in more barbaric times.

In North America, approximately 15% to 25% of women… were sexually abused when they were children. Most sexual abuse offenders are acquainted with their victims; approximately 30% are relatives of the child…

---

[1] Mel Gibson *Apocalypto* (2006)

> Most child sexual abuse is committed by men... The most-often reported form of incest is father-daughter and stepfather-daughter incest...[1]

The consequences on the spirit and psyche of a young girl who has been raped by a father-figure are extreme and devastating.

> Child rape can result in both short-term and long-term harm, including psychopathology in later life. Psychological, emotional, physical, and social effects include depression, post-traumatic stress disorder, anxiety, eating disorders, poor self-esteem, dissociative and anxiety disorders; general psychological distress and disorders such as somatization, neurosis, chronic pain, sexualized behavior, school/learning problems; and behavior problems including substance abuse, self-destructive behavior, animal cruelty, crime in adulthood and suicide... Long term negative effects on development leading to repeated or additional victimization in adulthood are also associated with child rape. The risk of harm is greater if the abuser is a relative...[2]

It is a phenomena of these last days that such sexual abuse has come into the open, by fathers, uncles, teachers, priests , etc.

> Most children are abused by someone they know and trust. A study in three states found 96% of reported rape survivors under age twelve knew the attacker. Four percent of the offenders were strangers, 20% were fathers, 16% were relatives and 50% were acquaintances or friends.[3]

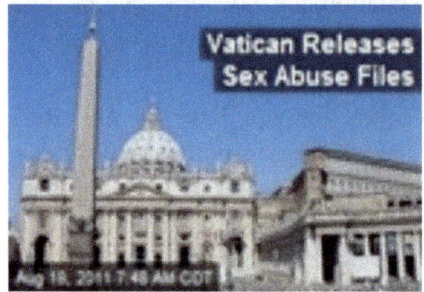

Vatican Releases Sex Abuse Files

Dysfunctional Humans

The abused becomes the abuser, and so it continues down the generations. The study of the science of epigenetics was initiated in the last decade by the astonishing discovery that women who had experienced the starvation and deprivation of the Siege of Stalingrad had grandchildren with significantly shorter life expectancies.[4] So it is not too much of a conceptual stretch to thing that the far more devastating experience of child rape by a parent figure would effect all of the descendants. This epigenetic imprint is the Original Sin, the twisted wisdom of the ancestors, which has kept all people from being able to love as God loves.

The merit of the age is such that father-daughter incest is universally deplored. Even so, it occurs in even the most spiritually-advanced communi-

---

[1] http://en.wikipedia.org/wiki/Child_sexual_abuse

[2] http://en.wikipedia.org/wiki/Child_sexual_abuse

[3] Advocates for Youth, 1995

[4] http://chd.ucsd.edu/seminar/documents/Morgan.08.pdf

ties.[1] Epigenetics explains the seemingly-unfair warning in the Bible: "Visiting the iniquity of the fathers upon the children and the children's children to the third and the fourth generation."[2]

A consequence of the Fall, the Bible recounts, is the prediction that Man would now have to earn a living by great effort. *Cursed is the ground for your sake; In toil you shall eat of it.*[3] For now there could be no natural servant class. The Neanderthals were not domesticated with God's Love; instead they were all completely exterminated, and even eaten,[4] by brutish, fallen humans within a few thousand years.

> [A] possibility is that competition for resources flared up into violence and genocide. Tolerance is not a Sapiens trademark. In modern times, a small difference in skin color, dialect or religion has been enough to prompt one group of Sapiens to set about exterminating another group.[5]

The human brain is organized into a hierarchy of modules that each arose in evolution as a step in the expression of the Logos. At the very bottom is the gut brain; the brain stem and the voluminous, if diffuse, network of ganglia that are spread over the internal organs. This is concerned with the survival of the self, and it arose in fish and was perfected in reptiles.

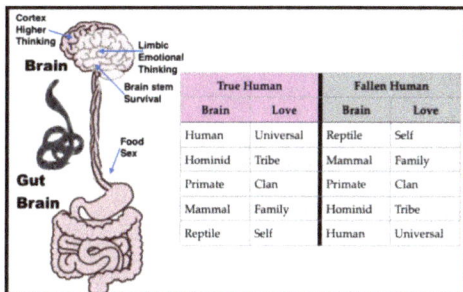

Brain		True Human		Fallen Human	
		Brain	Love	Brain	Love
Cortex / Higher Thinking		Human	Universal	Reptile	Self
Limbic / Emotional Thinking		Hominid	Tribe	Mammal	Family
Brain stem / Survival		Primate	Clan	Primate	Clan
Food / Sex		Mammal	Family	Hominid	Tribe
Gut Brain		Reptile	Self	Human	Universal

Above this is the mammalian brain with ability to love on a family level. Above this is the primate brain, with ability to love on the clan level. Above this is the Hominid brain, with ability to love on the tribal level. At the very top of the hierarchy, is the human module with the potential for universal love.

The trauma of the Fall reversed this hierarchy, and the self-centered Gut brain became dominant. Incidentally, much of the Gut brain is a hollow tube about the intestines, making the snake an appropriate symbol for Satan. This is why our Founder is always admonishing that: "Your body is your enemy." The human level and the gut level are at war for control of the whole.

---

[1] Livin' Life Ministry, Manhattan Center, July 28, 2011

[2] Exodus 34:6-8

[3] Genesis 3:17

[4] http://kasamaproject.org/2010/12/28/when-red-haired-neanderthals-were-eaten/

[5] Harari, Yuval Noah. Sapiens (pp. 17-18). Harper.

The cycle of dysfunction, the inability to love, has been passed on from generation to generation. The history of restoration is God's effort to lead humanity out of this dreadful state, to break the cycle. The cycle of dysfunction will be broken with the advent of the True Parents and the start of the original ideal.

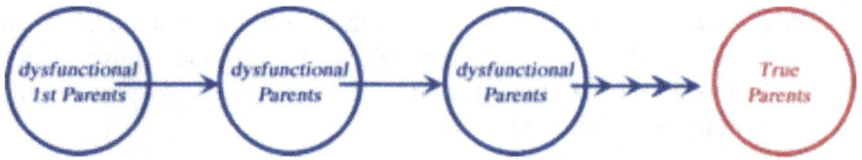

God's work of Restoration of this Ideal is constrained by the Logos, which is still running things, and human responsibility and ancestral merit, the opposite of sin.

This topic is beyond the scope of science at this time, and is more properly handled in a religious perspective.

9 • POSSIBILITIES FOR INTERSTELLAR TRAVEL

T his chapter is not so much about the merger of religion and science, but more of suggestions to explore the possibilities that religion posits for science to establish.

NECESSITY OF INTERSTELLAR MIGRATION

The necessity for a method for the human race to spread out to the stars and galaxies is driven by the mathematical properties of geometric growth. Anthropology suggests that the human First Ancestors appeared in the midst of a prehuman population ~100,000 years ago marked by the transition from the Paleolithic Age—with cycles of innovation taking millions of years—to the Neolithic Age with cycles taking just centuries, even decades. Genetic studies also place humanity's origin in the same time period with the studies of Mitochondrial Eve and Y-chromosome Adam, followed by the Out of Africa migration to populate eventually all land areas of the globe.

Population growth, however, was not geometric over the succeeding 80,000 years of Fallen history inasmuch as disease, war and famine kept the surviving human population small for most of Fallen history. The graph[1] on the right gives an estimate of human population growth showing that, for most of Fallen history the population was small. It is only in the last few centuries that the population has grown dramatically, passing the six billion mark as we entered the third millennium—an intimation of the power of geometric growth.

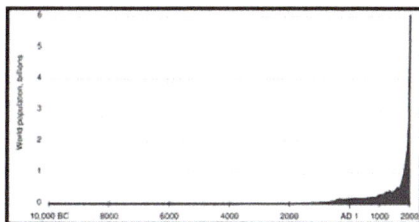

The *Divine Principle*, however, is most emphatic in stating that the Fall of the First Ancestors was not a part of the Original Plan. The Creator intended a humanity expressing True Love to emerge and the restraints of war, disease or famine on population growth to never emerge.

_____

[1] Katja Keuchenius *False Numbers: Muslims And Population Growth* http://www.united-academics.org/magazine/getting-right/muslim-global-population/ 2013

In the intended history, it is a reasonable assumption that the human population would grow exponentially. The current world population of 7.8 billion, with an annual growth rate of 2%, could have been reached in just 1,200 years!

In our time, 100,000 years after the start, the population would be: 10^{860} an enormous number of people, far beyond what any conceivable technology might enable the Earth to sustain with its surface area of only 5×10^{15} square feet. This problem was first publicized in 1798 by Thomas Malthus who pessimistically wrote (inspiring Darwin):

> Famine seems to be the last, the most dreadful resource of nature. The power of population is so superior to the power of the earth to produce subsistence for man, that premature death must in some shape or other visit the human race.[1]

Clearly, the Original Plan was either to invoke drastic birth control measures or to provide for expansion on a plethora of planets other than the Earth. As the first alternative is anathema to the spirit of the *Divine Principle* view of the Creator, we can expect provision was made for the second alternative—expansion to other planets. This view, unlike that of Malthus, is decidedly optimistic.

The *Divine Principle* asserts that this was supposed to have happened early in human history if the Fall had not occurred:

> External dominion [the Third Blessing] means domination through science. If man, having perfected himself, had been able to dominate the world of creation internally with heart-and-zeal identical to that which God had over the world of creation at the time of its creation, man's scientific achievement could have reached its culmination in an extremely short time, because man's spiritual sensibility would have been developed to the highest dimension. Thus, men could have dominated externally all the things of creation. In consequence, man not only could have subdued the world of nature, **including the heavenly bodies**, at the earliest possible date, but he also could have brought about an extremely comfortable living environment due to the economic development that would have accompanied scientific achievement.[2]

Many Earths

Fortunately supporting this perspective, modern astronomy has determined that there are plenty of Sun-like stars in the observable universe. Current estimates are that there are 100 billion planets in our galaxy[3] and an equal number in all the other galaxies.

---

[1] Malthus T.R. (1798) *An essay on the principle of population.* Chapter VII, p. 61

[2] *Divine Principle* p.128 (1977 edition)

[3] https://www.space.com/19103-milky-way-100-billion-planets.html

Some of these will orbit unsuitable suns—the O- and B- and A-type stars being too large and hot, while the M-type stars are too small and cool to be suitable. Some will be too large or too small, others will not be in the Habitable Zone about their sun where water exists as gas, liquid and solid. Nevertheless, in our galaxy alone, there are probably 50 billion planets suitable to humans.[1]

In our solar system, the Earth is smack in the center of the Habitable Zone of our G-type sun where temperatures are just right. Mars is outside the outer boundary of the zone and too cold with carbon dioxide a solid, while Venus is outside the inner boundary and too hot, where lead is a liquid. Slightly larger, hotter F-type stars, and slightly smaller, cooler red K-type stars also have more or less respectively expanded habitable zones about them.

The red region [in the diagram] is too warm, the blue region too cool, and the green region is just right for liquid water. Because it can be described in this way, sometimes it is also referred to as the "Goldilocks Zone".[2]

It is estimated that the observable universe contains at least another 100 billion galaxies like ours, each of which has its own abundance of amiable stars and Goldilocks Zones.

While the earliest theories of planet formation suggested that planets would be rare about stars, current theories imply that they are commonplace. This view is corroborated by the recent detection of Jupiter-like exoplanets orbiting dozens of the nearest stars:

Exoplanets have become one of the most exciting and important topics in astronomy today. In addition to finding over 5,000 new worlds, scientists using tools like NASA's Kepler mission have found that not only are exoplanets as plentiful as stars in our galaxy, but that a sizable portion of them are small, rocky planets like Earth. It's possible that, only a few years from now, astronomers will be able to find a habitable planet like our own orbiting another star.[3]

On July 23, 2015, NASA made the announcement of the first earth-like planet to be discovered:

[1] https://blog.physics-astronomy.com/2022/09/astronomers-admit-we-were-wrong100.html?fbclid=IwAR2dtNImujiA5UbTzQ9U9G03mwIcg9YXQTdOKFAB8izgB9jGLJLx2JOFC00

[2] *The Habitable Zone* https://www.e-education.psu.edu/astro801/content/l12_p4.html

[3] *20 YEARS OF EXOPLANETS* http://planetquest.jpl.nasa.gov/page/20-years

NASA's Kepler mission has confirmed the first near-Earth-size planet in the "habitable zone" around a sun-like star. This discovery and the introduction of 11 other new small habitable zone candidate planets mark another milestone in the journey to finding another "Earth."[1]

Intensive study of this topic has developed rapidly until there appear to be more earth-like planets in our galaxy than people on Earth:

Astronomers using NASA data have calculated for the first time that in our galaxy alone, there are at least 8.8 billion stars with Earth-size planets in the habitable temperature zone.[2]

It has been firmly established that simple life appeared on the Earth at the very beginnings of its 4.5-billion-year history—simple bacteria emerging about 4 billion years ago. As one Nobel Laureate affirms, this rapidity logically implies that life is a very probable occurrence:

What this ... implies with respect to the assembly of the first cell is that most of the steps involved must have had a very high likelihood of taking place under the prevailing conditions.... In other words... the universe was—and presumably still is—pregnant with life.[3]

As noted earlier, unlike conventional thinking, the Logos and the wavefunction make seemingly impossible things possible, and seemingly possible things impossible. They load the dice, so to speak.

While intelligent life, and even trees and rats, took much, much longer to get established than bacteria, this simple life rapidly transformed the atmosphere of Earth from its original poisonous anoxia into one of bountiful oxygen and inert nitrogen as early as 3.5 billion years ago. We can conclude from both science and theology that we would expect to find that there are plenty of habitable Earth-like planets in the habitable zone with an oxygen atmosphere scattered throughout the universe. One early estimate of planets with liquid water and an oxygen atmosphere calculated that our galaxy might contain 600 million such benevolent planets[4] ripe for human habitation.

Encountering non-Earth life, even bacteria, will have a profound impact on biological thinking. Current thought, based on the atheism that dominates modern science, is that life was accidental and that the biochemistry underlying life's functioning could have been radically different. The origin and evolution of living systems was utterly contingent and, as evolutionist S. J. Gould famously asserted in a thought experiment, would be entirely

---

[1] http://www.nasa.gov/press-release/nasa-kepler-mission-discovers-bigger-older-cousin-to-earth

[2] Seth Borenstein. *Science News*, Nov. 4, 2013.

[3] Duve, Christian de, *Vital Dust: Life as a Cosmic Imperative*. New York: Basic Books, 1995. p.9

[4] Asimov, Isaac, *Extraterrestrial Civilizations*, New York: Crown Publishers, p.169

different if the development of living systems was repeated over from the start:

> I call this experiment "replaying life's tape." You press the rewind button and, making sure you thoroughly erase everything that actually happened, go back to any time and place in the past . . . . Then let the tape run again and see if the repetition looks at all like the original.... any replay of the tape would lead evolution down a pathway radically different from the road actually taken.[1]

The theistic view, as exemplified in the *Divine Principle*, takes the opposite point of view. The structure and functioning of living things was charted out in the Original Plan. We can expect life on an exoplanet to take a very similar path to that on Earth. We can expect, for example, that life will involve L-amino-acids in proteins—the masters of analog manipulation—and D-nucleotides in DNA/RNA—the masters of digital manipulations. We do not expect to find life using D-amino-acids or L-nucleotides which are toxic to our kind of life.

Such a showdown between a contingent view and a created view of biological evolution is a long way off, but astronomers have already found planets that could harbor simple life:

> What about Earth-like planets with Earth-like orbits? Of the 461 new planet candidates, 51 of them are in the so-called "habitable zone," the Goldilocks region around the star that's at just the right temperature for liquid water to exist. And one of these new planet candidates has all three of the qualities we're looking for in a twin Earth: it's in the habitable zone, it's only 1.5 times the size of Earth, and it's orbiting a sun-like main sequence star.[2]

Finding such planets with a telescope is relatively easy. The real challenge is getting there from Earth so that we can explore, and hopefully populate, those exoplanets. Religion says that there is plenty of room out there, astronomers tell us there are many planets out there. The final challenge for science it is, How do we get there? First, the classical reasons why it's impossible; second, the optimistic view of modern science.

The Classical Challenge

Classical physics affirms the commonsense view that in order to get from point A to point B you need to traverse all the points that lie between them. This holds for material things as well as intangibles such as information.

---

[1] S. J. Gould, *Wonderful Life: The Burgess Shale and the Nature of History*, New York: Norton, 1989, p. 48-51

[2] Moyer, Michael, *Earth-Like Planets Fill the Galaxy*, 2013 http://blogs.scientificamerican.com/observations/2013/01/08/earth-like-planets-fill-the-galaxy/

The first problem is that even the nearest stars are very, very distant. Alpha Centauri is considered a nearby star but is 40 trillion kilometers away. It takes speedy light more than 4 years to get there from here—traversing a distance of 4.4 light years—while our Home Galaxy is 100,000 light years across and the nearest galaxy, Andromeda, is 4,000,000 light years distant.

Current technology is only capable of attaining speeds much less than the speed of light, and theory suggests that velocities greater than 10% light's speed (~70 million mph) would be fatal since interstellar gas would, at such speeds, be encountered as lethal, high-energy cosmic rays almost impossible to shield against.

Forty years' travel to the nearest star is not an inducement to human migration. The centuries it would take to reach the majority of stars in our galaxy would entail a multigenerational journey—providing great scenarios for science fiction writers but not a convenient way of relieving population pressure on Earth. For example, the recently discovered planet Kepler-452b (Earth-2) is 1,800 lightyears distant. At 10% light speed, it would take 18,000 years to get there!

The second major problem is that the vacuum of space is utterly hostile to life—no air, no water, no food, no atmospheric shielding of cosmic rays— so hostile that even simple errors can lead to rapid death. A detailed compilation of the hazards encountered in space travel has been published.[1]

For these reasons we conclude that the Creator must have had a quite different method of travel in mind for His ever-expanding family. For clues to what this might be, we look to the confluence of theology and science in the *Divine Principle* and quantum physics.

Dual Characteristics

The Principle of Creation states that all things are created with two unified sets of dual characteristics. The primary duality is the vertical unity of internal character (mind) and external form (body). The secondary duality is the horizontal interaction between male and female animals and plants, electropositive and electronegative chemical elements, and plus and minus electric charge.

To summarize what was discussed earlier: Quantum physics affirms the same basic principle, albeit using different terms and mathematical precision. The fundamental entities which interact to form atoms, molecules, cells and all things have a dual nature. They have an internal wave aspect which is mathematically described by complex numbers (combining linear size and

---

[1] Comins, N.F. 2007, *The Hazards of Space Travel*, New York: Villard Books.

circular rotation), and an external particle aspect which is mathematically described by regular numbers (with linear size only).

The mathematical connection between the two types of numbers is simple: the absolute square of the internal complex number generates the external real number which gives the probability of what the external particle will be and do in space and time. In classical physics, the probability of an event is calculated by adding up the probability for all the different ways in which the event can occur. In quantum physics, it is the probability amplitudes—the name given the complex number—that must be added, then squared to give the final, real probability. It is the great difference between adding real numbers and adding complex numbers that is the source of most of the "weirdness" that classical physicists attribute to quantum science.

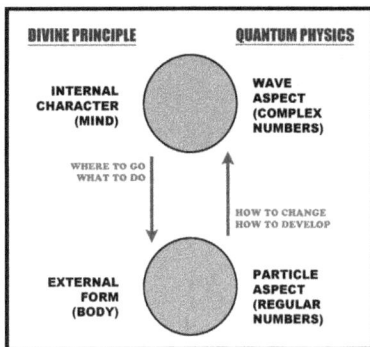

Even at the most basic levels of the physical world, such as in the behavior of electrons and photons, however, the particle has the ability to freely choose its future within the bounds of the probability. So, if the internally-generated probability is 0% or 100%, there is no freedom of choice; if the probability is divided 50-50 between two paths, there is no way to predict which of the two paths will be taken—the choice is totally random and unpredictable.

The internal wave is non-local and spread out in spacetime while the external particle is localized at a position in spacetime. Heisenberg's Uncertainty Principle prevents this from being at a point—with zero extension—so it jitters about, but with high-energy probes the electron has been located to within a billion, billionth (10^{-18}) of a meter,, which is a tiny locality even on an atomic scale.

The internal and external aspects, as expected from the *Divine Principle* perspective, are unified and reciprocally related to each other. Quantum physics states that:

a. The internal wave determines the probability of where the external particle will be in spacetime and what it will do there.

b. The external interactions of the particle determine how the internal wave will change and develop in space and time.

Both disciplines agree that the internal wave aspect is subjective, while the external particle aspect is responsive. This is the source of much confusion to scientists with a classical perspective since the external particle always does what the internal wave tells it to do, even if the instruction is to do something quite impossible in classical physics (which is, of course, ignorant of the internal aspect to reality).

One of the most classically impossible behaviors involves the nodes of waves, i.e., places where the wave is zero and where the particle can never be.

Wave Nodes

Any wave, such as a sine wave, has an amplitude (shown in blue in diagram) that goes from positive to negative and back again. The square of the amplitude is the intensity of the wave (shown in red) and is always positive. For the internal wave aspect in quantum physics, it is this intensity that generates the always-positive probability that directs the behavior of the external particle.

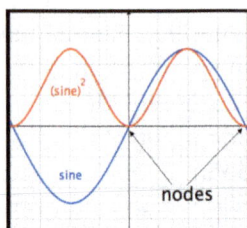

Where the internal amplitude crosses zero from positive to negative or vice versa, the external probability is exactly zero—it is impossible for the particle to be at that location.

It is such nodes that lead to what classical scientists call "quantum weirdness." For example, the standing electron waves that occur in the atom can have two lobes, such as in the 2p orbital, that are separated by a node at the atomic nucleus. The electron particle spends 50% of the time in one lobe and 50% in the other, but spends zero time at the node that separates them. In the vernacular, we would say it teleports between the two lobes, ignoring the space that separates them, while the technical term is tunneling between the two lobes.

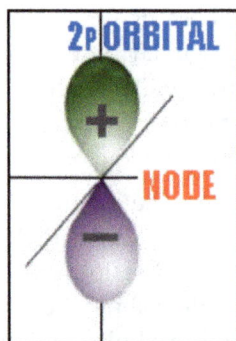

Such behavior is utterly nonsensical in classical physics—where the negative electron would sit on the positive nucleus, not avoid it completely—but makes sense in quantum physics where the wave determines what the particle will do.

The atomic node here only spans a few nanometers, but even stranger behavior occurs when the node stretches over a considerable spatial separation, as it does in the slit experiment which played such a leading role in the revolution from classical to modern physics.

SLIT EXPERIMENT

In a suitably designed apparatus, the wave aspect can be can be split into two and the node that separates the two lobes can cover many centimeters. This is demonstrated by the simple slit experiment where the wave is separated into two lobes that pass separately through two slits in a barrier. On the far side, the two waves interfere with each other creating fringes of high probability and low probability—and this occurs even when a single particle at a time passes through the apparatus.

To the classical mind, it appears that the single particle passes through both slits at the same time. This is not so. Rather, the particle spends 50% of the time passing through one slit, and 50% in the other, tunneling rapidly from one location to the other. Attempts to detect which slit the particle is traveling through inevitably involve an interaction, and this inevitably alters the internal wave and changes the result, with a loss of interference.

Such slit experiments dividing the internal wave have been performed on simple things—such as electrons, photons and single atoms—as well as complex molecules such as fullerenes with sixty or more atoms in their spherical structure. This molecule is definitely 'matter' exhibiting the reality of its internal wave aspect. It is probably only a matter of time until some brilliant experimenter manipulates the internal wave of a bacterium and gets its external body to seemingly pass 'through both slits at the same time.'

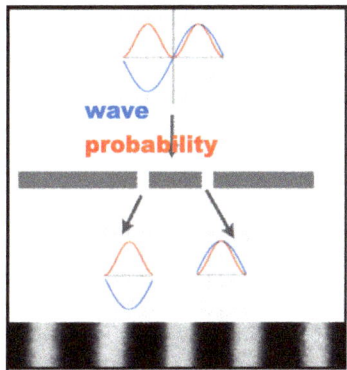

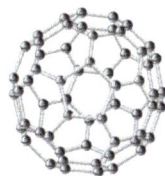

The Born-Oppenheimer approximation in quantum mechanics treats nuclei as practically stationary with respect to the electrons. But distortions in the orbitals caused by external interaction are often accompanied by changes in the molecular structure, i.e., the positions of the nuclei. This change in structure is particularly important in protein activity where interaction with, for example, a calcium ion can radically alter the protein's wavefunction which alters the folded structure and activity—a conformation change.

Entanglement

This disdain in quantum science for the classical rules of how material entities are supposed to behave and how spatial separation is to be respected is even more in evidence in the phenomenon of entanglement. This behavior is so anathema to the classical mind that even the Nobel Laureate who first theoretically noticed its possibility—Albert Einstein, no less—vehemently rejected it as "spooky."

> Entanglement occurs when two particles are so deeply linked that they share the same existence. In the language of quantum mechanics, they are described by the same mathematical relation known as a wavefunction. Entanglement arises naturally when two particles are created at the same point and instant in space, for example. Entangled particles can become widely separated in space. But even so, the mathematics implies that a measurement [interaction] on one immediately influences the other, regardless of the distance between them. Einstein and company pointed out that according to special relativity, this was impossible and therefore, quantum mechanics must be wrong, or at least incomplete.   Einstein famously called it 'spooky action at a distance.'[1]

Entanglement has been observed for many properties of many entities—such as right/left rotation for photons, $\pm 1/2$ spin for electrons, up & down magnetism for protons, etc. Einstein tried to explain this phenomenon by postulating hidden local variables. In his attempt to avoid accepting a nonlocal aspect to matter, Einstein declared that this correlation was a result of a hidden, local variable carried by the entangled entities. This hidden variable would be classical and be measured by a real number associated with each photon. In 1935, along with two other authors, he published a paper claiming that Quantum Mechanics was incomplete, and made nonsensical predictions, a viewpoint known ever since as the EPR Paradox.

> It is an early and influential critique leveled against ... quantum mechanics. Albert Einstein and his colleagues Boris Podolsky and Nathan Rosen (known collectively as EPR) designed a thought experiment which revealed that the accepted formulation of quantum mechanics had a consequence which had not previously been noticed, but which looked unreasonable at the time. The scenario described involved the phenomenon that is now known as quantum entanglement.[2]

In 1964, a way of differentiating between local and a non-local correlation was formulated, known as Bell's Inequality.[3] The math is basically simple (if often shrouded in sophisticated symbolism) and involves not having the detectors at identical orientations. This reduces the correlation between the results at either end.

If the correlation involved local hidden variables (that could not influence each other once separated), the alterations would sum together as real

[1] *Einstein's "Spooky Action at a Distance" Paradox Older Than Thought*, 2012
http://www.technologyreview.com/view/427174/einsteins-spooky-action-at-a-distance-paradox-older-than-thought/

[2] *EPR paradox* https://en.wikipedia.org/wiki/EPR_paradox

[3] Bell, John (1964). "On the Einstein Podolsky Rosen Paradox" *Physics* 1 (3): 195–200.

numbers. If the connection was non-local, they would not, and they would sum up as complex numbers do.

An illustration of how different these are: For two regular numbers with a linear magnitude of 2, adding them together always gives a regular number with magnitude 4 or 0. For two complex numbers with magnitude 2, however, adding them together can result in a complex number with a magnitude of anything from 0 to 4 depending on their relative rotations.

Experiments to test this inequality also had to avoid any possibility of the detectors somehow passing information about their orientations at light speed. If the distance between the two detectors was such that it took 10 nanoseconds for light to cross it, the settings of the detectors had to be altered in 1 nanosecond to prevent any connection at the speed of light.

These experiments have been performed—the most unusual involved kilometers of optic cables threaded through the sewer system of Vienna!—and they have all proved unequivocally that the entangled connection is non-local, instantaneous and independent of the spatial separation. [The concepts briefly mentioned in this section are thoroughly discussed in the book, *The Age of Entanglement*.[1]]

Intergalactic connections

While the indifference of the internal aspect of matter to spatial separation—as illustrated by nodes and entanglement—has been experimentally observed over the range of nanometers to kilometers, there is no theoretical limit to how great the ignored spatial separation can be. This is the foundation for a possible means of interstellar and intergalactic travel.

There are many natural situations in which a pair of entangled entities is emitted. One well-studied example is the calcium atom which, when excited by a thermal collision or absorption of a photon, can revert to the ground state by emitting a pair of entangled photons that zip off in opposite directions while remaining in the same wavefunction.

Calcium atoms floating in space regularly emit such pairs of photons. As mentioned earlier, the internal wave determines what the external particle does, while the external interactions of the particle determine how the wave alters. Unless special care is taken, if either particle interacts externally, the internal wave alters and the entangled state is lost—the phenomenon of decoherence.

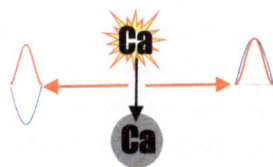

---

[1] Gilder, Louisa, *The Age of Entanglement: When Quantum Physics was Reborn,* New York: A. A. Knopf, 2008

Luckily, converting what we earlier considered a negative point into a positive, interstellar and intergalactic space are essentially empty, so it is quite possible for these photons to travel for millions, even billions, of years without either of them interacting, while yet retaining the entangled state for millions or billions of years as their spatial separation ever increases.

The star Alpha Centauri is 4 light years distant from us. If 2 years ago a calcium atom directly between us emitted a photon pair, and one of them reached the Earth, the other would be in the vicinity of Alpha Centauri—a non-local connection between here and there. If a similar thing happened 200 million years ago between here and the Andromeda Galaxy, we would have a non-local connection between the two galaxies.

If the entangled pairs were generated in the last 500 million years, the non-local connections would be spread out in a sphere of with a diameter of 1 billion light years. It is only in the last few decades that techniques have been developed that allow the contents of this sphere to be mapped in some detail.

In this sphere are ~250 quadrillion stars, in ~3 million galaxies comparable to the Milky Way, associated into ~100 galaxy superclusters, all of which have been given names. (Our Milky Way Home Galaxy is on the fringes of the Virgo Supercluster at the center of the sphere.) A few of these are shown in the following diagram of a 2-D slice of the universe.[1] It does seem crowded with galaxies, but all are so far away that only the very

Our local supercluster of galaxies ("Laniakea") Every point of light in this image represents an

nearest galaxy, Andromeda, is visible to the naked eye; all the rest require sophisticated telescopes to be observed.

OBSERVABLE AND UNOBSERVABLE

All the stars and galaxies in this vast sphere are essentially the same as our galaxy, billions of stars orbiting around a central quiescent black hole of millions, even billions, of solar masses.

As we look further and further into the depths of the universe, however, things start to look very different. At distances over 3 billion light years we start to observe quasars with active black holes at the center, so active that they are visible even at a distance of 13 billion light years. If the Milky

---

[1] *The Universe within 1 billion Light Years: The Neighboring Superclusters* http://www.atlasofthe-universe.com/superc.html

Way had such a violent center, life even at our 30,000-light-year distance from the center, would be impossible, the intense radiation would sterilize the earth.

This might seem to set a limit to human intergalactic expansion. Fortunately, contemporary quasars are an artifact of the slowness of light on the scale of superclusters. For we are seeing those far distant galaxies in their formation stage, when the central black hole was forming and clearing out its immediate neighborhood of stars and gas. An observer 10 billion light years distant looking our way would see the Milky Way as a quasar as our central black hole was vacuuming up everything in its vicinity. This youthful exuberance ended billions of years ago, and our central black hole has settled into its amenable middle age.

All those distant galaxies have also settled into middle age, and are no longer in their boisterous youth. Every single galaxy in the entire physical universe is roughly the same age as the Milky Way, They all originated in the gravitational condensation of material generated at the moment of Creation, the Big-Bang that occurred ~13.5 billion years ago.

One of the few aspects of science history that almost everyone is aware of is that in ancient days everyone—including the Biblical authors—assumed that the Earth was at the center of the universe, and that everything else rotated about the earth.

One of the first successes of modern astronomy was to displace the Earth from the center of things, and replace it with planets orbiting the Sun. This Copernican Revolution in the 1500s was the shift from the Ptolemaic geocentric model to a heliocentric model with the Sun at the center of the Solar System.

As the adage has it: What goes around, comes around. This is exemplified by modern cosmology which has replaced the parochial geocentrism of Ptolemy with a cosmic geocentrism that places the Earth at the exact center of the Observable Universe. Again this is due to the tardiness of light of light on the cosmic scale. For in a universe that is only 13.5 billion years old, the most distant things we can observe are those whose light can reach us in that time period. Anything more distant cannot be observed.

The distance of 13.4 billion light years defines the boundary of the Observable Universe, a perfect sphere with us at the very center. The surface of this sphere is created by the Recombination Era—the time when the expanding universe had cooled to well below the surface temperature of the Sun and the plasma of free electrical entities (as found inside the Sun and neon tubes) could condense into neutral atoms and the light could travel unimpeded through empty space. This light has been stretched nowadays into the

Cosmic Microwave Background, and beyond this barrier nothing is observable in any variety of light.

This boundary to the Observable Universe is uniform to 1 part in 10,000. At greater resolution, however, speckles of slightly hotter or cooler temperatures are observed—shown in orange and blue respectively in the following photo of the boundary to the Observable Universe taken recently by the European Space Agency's PLANCK Mission.[1]

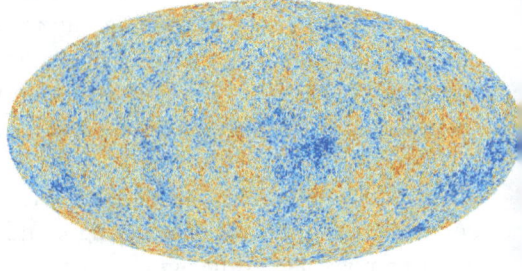

The rest of the physical universe, the Unobservable Universe, is also the same age as the Observable Universe, and contains a plethora of galaxies of stars and planets just waiting to be explored. Estimates of how much larger this unobservable universe is compared to the observable universe vary from a moderate 250 times larger[2] to a mind-boggling 300 billion trillion times larger[3]. Given the power of geometric growth, I would suggest that God's Original Plan embraces the larger, rather than the smaller, estimate.

Luckily for our descendants who will be looking for planets to populate, the seeming barrier between these two parts of the universe is utterly transparent to the entangled non-local connection we are discussing. An entangled pair of photons created 10 billion years ago would be separated spatially by 20 billion light years. One entering our solar system would have its twin well inside the unobservable universe, and provide a non-local connection between the two.

Role of the Moon

As hitting the earth's atmosphere and interacting with a gas molecule would cause decoherence, the best place to harvest these interstellar, intergalactic and inter-universe connections would be on the airless Moon—a great example of the Creator's planning ahead. Not that the Moon has not already proved itself invaluable, even essential, over the eons.[4]

---

[1] http://www.esa.int/spaceinimages/Images/2013/03/Planck_CMB

[2] Vanessa D'Amico *Universe Could be 250 Times Bigger Than What is Observable* 2011 http://www.universetoday.com/83167/universe-could-be-250-times-bigger-than-what-is-observable/

[3] Alan H. Guth. *The Inflationary Universe: The Quest For A New Theory Of Cosmic Origins*. New York: Basic Books, 1997. p. 186.

[4] Comins Neil F.,*What If the Moon Didn't Exist?: Voyages to Earths That Might Have Been*, New York: Harpercollins, 1993

With the insight provided by quantum physics, we find that our Moon is constantly bombarded with a plethora of non-local connections to locations scattered over the entire physical universe. This is certainly the kind of situation that the *Divine Principle* leads us to expect. A caveat should be noted at this point: Even though these entangled connections uncovered by 20th century physics seem to provide the expected interstellar connections, the beneficent Creator might have had quite a different travel plan in mind, intimations of which are unknown to current science.

FUTURE DEVELOPMENTS

It is disappointing, I know, because this is about as far as we can go, since Entanglement science and technology are only in their infancy. I would be delighted to expound on how entanglement-created Intergalactic Connections are physically related to Interstellar Travel. But this is the work of the next few centuries—all we know is that God must have planned for intergalactic travel, and that we know of a natural source of intergalactic connections. Putting the rest of the picture together is the task for the technologists of generations to come.

This formation stage of Entanglement Science can be likened to that of Electromagnetic Science and technology around 400 years ago when Luigi Galvani was twitching frog muscle with silver and copper wires in the late 1700s, and Michael Faraday was waving wires and magnets at each other in the early 1800s. On such simple foundations was built the Age of Electricity in just three centuries. (I can testify to the cultural importance of electricity, having personally experienced the Great Blackout of New York City in 2003.)

Current understanding has no clue, for instance, as to how to manipulate such non-local connections for communication, let alone bodily travel. Trapping entangled particles without decoherence is also a problem. Progress is being made, however, such as moving beyond entangled pairs to entangled multitudes:

> The largest number of particles that has been entangled so far is four. However, the Innsbruck-Aarhus team claim that their [Bose-Einstein condensate] technique could eventually be used to entangle any number of atoms.[1]

One problem is that most scientists are essentially materialists—they feel uncomfortable about including an internal aspect in their speculations about what is possible. So disagreeable is this internal aspect that most scientists try to ignore it as soon as possible. The internal aspect is dominant in physics, useful in chemistry, and barely mentioned in basic biochemistry

---

[1] http://physicsworld.com/cws/article/news/2000/jun/29/quantum-entanglement-spreads-to-bose-condensates

(where the prevailing model is the external fitting together of lock and key). The internal aspect of matter, however, is essentially absent from biology, genetics and neuroscience; it is totally ignored as if it did not exist. A scientist familiar with Unificationism, however, is quite comfortable with the concept of a subjective internal aspect and is equipped to go where others dare not.

The coming generations of Unificationist scientists will develop a flourishing science and technology based upon unified dual characteristics, and I hope will be pioneers of interstellar travel so that humanity can populate the Universe without limit.

It would have been impossible, even for a practical genius such as Michael Faraday, to take his contemporary understanding of electricity in the 1800s and predict the technology of GPS navigation, iPhones and the international Internet. Similarly, it is impossible to accurately predict what the Age of Entanglement will provide in the way of interstellar transportation. But that does not mean that we cannot speculate with the little that is already understood, creating the following blend of known science and science fiction.

The Future

Current research into entanglement is restrained by the constant struggle to avoid decoherence. The atmosphere about us is packed with molecules darting around with thermal energy, ready and able to interact at any moment and destroy the entangled state. To avoid this, experiments have to be performed in small leakproof apparatuses, where an expensive high vacuum can be maintained, and usually at liquid nitrogen temperatures or lower.

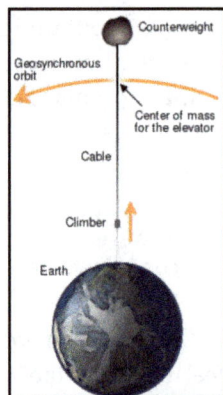

On the Moon things are quite the opposite. Vacuum and low temperatures are the norm so experiments can be performed in the vast open space, and it is the scientists who are maintained inside leakproof accommodations kept at room temperature. In other words, the perfect place for the science of entanglement to make rapid advances is a Moon base. As already mentioned, it is the only convenient place to collect naturally-entangled cosmic rays.

Even though the Moon, on an intergalactic scale, is right next door to us here on Earth, traveling to and from it using conventional chemical rockets is costly and fraught with problems. This inconvenience would be obviated by a *space elevator*, schematically illustrated on the right.[1]

---

[1] Diagram at: http://www.newworldencyclopedia.org/entry/Space_elevator

A space elevator is essentially a long cable extending from our planet's surface into space with its center of mass at geostationary Earth orbit, 35,786 km in altitude. Electric elevators traveling along the cable could serve as a mass transportation system for moving people, payloads, and power between Earth and space.

> Four to six "elevator tracks" would extend up the sides of the tower and cable structure going to platforms at different levels. These tracks would allow electromagnetic vehicles to travel at speeds reaching thousands of kilometers-per-hour.[1]

The space elevator is an idea that has been around for decades and only needs one technological advance to become feasible—a material that is lightweight yet 50 times as strong as steel. Diamond would do nicely, but nobody has come up of an inexpensive way of making synthetic diamond in the huge quantities that would be required. Cables of defect-free graphene—a distant cousin of diamond that is already being synthesized—might be up task:

> Graphene is the strongest material in the world, according to new experiments done by researchers at Columbia University in the US. The secret to the material's extraordinary strength, says the team, lies in the robustness of the covalent carbon-carbon bond and the fact that the graphene monolayers tested were defect-free.[2]

It is, perhaps, thus only a matter of time before cables capable of handling the space elevator are available and the Moon becomes accessible. So, for the first step in our future speculation, we postulate a space elevator facilitating inexpensive travel between Earth and Moon. Following this, establishment of a science facility on the Moon focusing on studies of manmade and natural entanglement. We will allow a century to pass until the technological manipulation of non-local connections becomes established.

The scientific foundation on the Moon then switches to the harvesting of the enumerable entangled particles that arrive every second, opening the non-local connection and seeing where the other end of the connection is located. Some of the recently created pairs will have their outer ends near the planets—a much better way to explore Mars than spending years in chemical rockets.

Most outer ends will be found, statistically speaking, in the great empty voids of space, either galactic or intergalactic. Amidst such dross will be the occasional jewel—a connection in the vicinity of another star. Telescopic

---

[1] NASA *Audacious & Outrageous: Space Elevators, 2000*
 http://science.nasa.gov/science-news/science-at-nasa/2000/ast07sep_1/

[2] *Graphene has record-breaking strength,* Jul 17, 2008
 http://physicsworld.com/cws/article/news/2008/jul/17/graphene-has-record-breaking-strength

examination through the non-local connection will determine if there are any potentially habitable planets. If there are, the connection will be opened wide and exploratory vehicles sent to investigate.

When a suitable planet is found, migration of families—hopefully true families—will commence. Once the essentials of civilization have been taken care of, a facility for harvesting the local showers of entangled particles will be created.

After a few centuries, a web of non-local connections between the stars and galaxies will be established. Such a non-local web will be traversed not by spaceships but subway cars, and humanity will spread God's kingdom of true love to the stars.

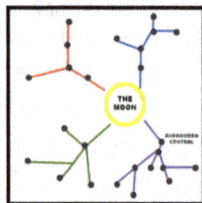

EPILOGUE

Science has established that the physical universe has an 'expiration date' inherent in its structure—luckily many tens of billions of years in the future. This eventual end is, basically, because the energy of the stars that warm planets and nourish plant life is derived from fusing the primordial hydrogen created in the Big-Bang into helium over time. For instance, our Sun gets its energy by converting, every second, ~600 million tons of hydrogen into ~596 million tons of helium and generating ~4 million tons of energy in its core, which then makes its way slowly—taking about a million years—to the surface where it is radiated out into space, a tiny fraction of which warms and nourishes us here on Earth.

In 10 billion years or so, our Sun will have consumed all its natal hydrogen and will shift to converting helium into carbon and oxygen. Unfortunately for our descendants, this will convert our Sun into a Red Giant with a diameter that will engulf Venus. Earth's oceans will boil away and become unlivable unless something drastic is accomplished.

One possible solution would be to open a truly gargantuan non-local connection and pass the Earth through it to an orbit around a young star. There are actually great clouds of tenuous hydrogen scattered throughout the galaxies, and new stars are being created out of these at a rate of about 10 per year. While the supply of this hydrogen is enormous, it is not infinite. When all the hydrogen is used up—and this will occur everywhere at about the same time—this physical universe will no longer be suitable for our kind of life.

We can assume that God has taken this into account, and that provision has been made for new and fresh universe to be created by humanity. The technology to generate a Big-Bang will be developed—if Dad can do something, so can His children. But all this is so far beyond our current understanding that we would do well here to suspend any further speculation. In preparation, however, we will take a look at Black Holes, and their role in Creation.

10. Big-Bang & Black-Holes

Unification Thought (UT) has the central belief that everything that God created has a role to play in the workings of the Logos and in fulfilling the Ideal of Creation involving His children, human beings. In this book we will explore how this perspective of UT can expand the scientific perspective.

Both religion and science deal with beginnings and endings. In the Bible's *Revelations*, God declares that, "I am the Alpha and the Omega, the Beginning and the End." Science also has a beginning to the universe—the Big Bang—and various scenario's of its ending. This chapter deals with the science aspect and speculates on a happy ending!

Metamorphosing spacetime

In the early days of science, matter was thought to be made up of little balls of stuff; and it was very difficult to see how bits of matter could just pop into existence out of nothing (without reintroducing the magic wand), because everyone thought that nothing and something were so very, very different.

While matter was considered to be a 'something' empty space and time were considered to be a 'nothing'. Something out of nothing, however, is no longer a problem for modern science, because scientists have discovered that something and nothing are actually very similar, and one can change into the other very easily.

An excellent and accessible overview of how science has solved the problem of something from nothing is *The Creation of Matter* by Harald Fritzsch. He offers an illuminating analogy: Take an extraordinary oven that can be set to any temperature you wish. Remove every single atom from the oven so you have absolutely no material inside. Our challenge is to see how we can make some matter out of this nothingness inside the oven. This will have to be, as we shall see, an ideal oven that can withstand any temperature and any pressure.

Even before turning on the oven, the inside is not quite empty, as there are a few particles (photons) of very low-energy electromagnetic radiation (infrared waves) zipping about inside, this is the thermal radiation at room temperature. But light is not really considered to be matter, and this is as empty as one can get.

Turn the oven on. As the temperature rises, two things happen: The particles of light get more energetic and there are lots more of them—16 times more for every time the temperature doubles. When the temperature of the oven has reached a few thousand degrees, the radiation is energetic enough to be perceived as light, and the inside of oven is brightly glowing a cherry red. As the temperature rises, green photons appear and the glow is now yellow like the sun's surface appears. Further up still, and blue is added so the oven's innards are now blindingly white.

A technical point is that temperature can be measured in °K (degrees Celsius above Absolute Zero) or in the energy of the thermal radiation measured in electron volts, eV, and we will use both, with the conversion factor being $1eV \approx 10^4$ °K.

By the time the temperature reaches millions of degrees—the temperature deep inside the sun—the radiation is mainly intense X-ray's and gamma rays—lots of them. Looking into the oven now would be fatal even from a distance of tens of miles! So far we have created only light, which is not really "something" in the material sense.

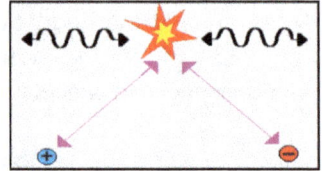

entity	photon E
± electrons	1 MeV
± muons	211 MeV
± protons	1.88 GeV
± neutrons	1.89 Gev
±C quark	2.5 GeV
± tauon	3.5 GeV
± B quark	8.5 GeV
Z boson	95 GeV
±W boson	160 GeV
±T quark	340 GeV

When the temperature gets up to six billion degrees—a temperature only reached in the universe these days for a few minutes during a supernova, the explosive death of a massive star—something quite surprising happens. Some of the high-energy gamma rays spontaneously turn into electrons and anti-electrons (positrons), a process called pair production.

This is a real case of "matter out of no matter," as the electron is one of the three ingredients of everyday material. The matter-antimatter annihilated back into photons as fast as it was produced as a steady state with an equal amount of photons and electron-positron pairs present. So, just by raising the temperature, we have filled our originally empty oven with huge numbers of photons, electrons, and anti-electrons.

As this phenomenon is usually described in reverse, by a falling temperature, the 1 MeV temperature is called the 'freezing point' of electrons and positrons; above this temperature they are as abundant as, and in equilibrium with, the photons; below it they annihilate and fall out of equilibrium as they combine into photons which are no longer energetic enough to create particle pairs.

Continue to stoke this miraculously heat-resistant oven. When the temperature reaches 10 trillion degrees another type of pair production begins: Protons, antiprotons, neutrons, and antineutrons are born. Protons and neutrons are the other two constituents of matter, so now our oven is filled with all we need. There are now so many particles in our originally empty oven that the density is a hundred million million, million kilograms per liter. This is "something out of nothing" with a vengeance! The charts list the temperature at which new entities join the mix and how the density of particles increases with temperature.

Temp	density= nucleons/litre
$10^{10}\,\text{K}°$	10^{31}
$10^{14}\,\text{K}°$	10^{47}
$10^{15}\,\text{K}°$	10^{51}
$10^{29}\,\text{K}°$	10^{107}

Naturally the pressure inside the oven is tremendous, and if the oven bursts there will be a tremendous explosion, and the contents will rapidly cool off. Can we expect to have "baked" a few quintillion tons of matter out of nothing?

Up until recently, the answer to this question was "no" because, while our hot oven is chock full of matter, it is just as full of antimatter. And matter and antimatter annihilate each other on contact, turning back into light. In our ultra-hot oven both are forming just as rapidly as they are disappearing. But if the oven cools, then the annihilation predominates, and by the time we are ready to open the door to see what we have baked, there is absolutely nothing left at all. Not one single atom. Just infrared photons as we started with. This seems a poor recipe for creating matter.

So, on solid experimental evidence it appeared that the Big Bang should have produced equal amounts of matter and antimatter. But the stuff of our universe is entirely matter, while antimatter remains a laboratory curiosity or an exotic fuel in science fiction dramas.

Yet we did create matter. We did bake something out of nothing even if it did end up flatter than a failed soufflé. What was lacking in our recipe for baking a universe? The answer is not to be found in cosmology but at the frontiers of modem physics, dealing with energies so immense that they are forever—well, for a very long time anyway—beyond the reach of experimental testing. The only time such energies have appeared in this universe was during the first instant of creation. This instant is the only "experimental set-up" on which to test these theories. *Unification Thought* asserts that there is an unknown aspect to the Logos that favors matter over antimatter, a current topic of intense research.

The Big Bang should have created equal amounts of matter and antimatter in the early universe. But today, everything we see from the smallest life forms on Earth to the largest stellar objects is made almost entirely of matter. Comparatively, there is not much antimatter to be found. Something must have happened to tip the balance. One of the greatest challenges in physics is to figure out what happened to the antimatter, or why we see an asymmetry between matter and antimatter.[1]

The Big Bang

The science of cosmology, the history of the entire universe, has established that our expanding universe originated in the Big Bang about 13.5 billion years ago. The current understanding of the Big Bang has progressed in tandem with high-energy particle physics, the behavior of matter at very high energies. This is because the universe started off in an ultra-high temperature and cooled as it expanded. This, of course, is exactly the reverse of our Miraculous Oven, which went from cool to very hot. But the sequence is remarkably similar.

Current experiments have reached TeV (10^{12} eV) levels, so our understanding of the early universe at such temperatures is well-established. The earlier times, and higher temperatures, are more speculative and have to depend on interpreting the relics that remain around for study. Examples of such relics are the comic microwave background photons, the ratio of primordial hydrogen/helium/lithium, and the very flatness of the vacuum of space.

The following is a summary of the current understanding of the Big Bang:

In the period from time zero to one Planck tick (10^{-44} second) a Planck volume (10^{-99} cubic meter) of False Vacuum emerged with a Planck mass (10^{-8} kilogram) at the Planck Temperature (10^{32} K, 10^{19} GeV). In *Unification Thought*, this is a consequence of God setting the Logos in action. All the developments that followed were encoded in the Logos.

This False Vacuum had an enormous negative pressure and entered a period of exponential inflation in which the Planck volume doubled in size each Planck tick. This was the creation of spacetime. The speck of False Vacuum expanded enormously in the Abstract Realm. The initial contents of the speck contained every possible entity, including quarks, and these were diluted as their separations increased exponentially.

Now, while the energy most forces fall off with separation, the chromodynamic energy does not, it *increases* exponentially with

entity	photon E	Big-Bang
± electrons	1 MeV	
± muons	211 MeV	
± protons	1.88 GeV	
± neutrons	1.89 Gev	
±C quark	2.5 GeV	
± tauon	3.5 GeV	freeze out
± B quark	8.5 GeV	
Z boson	95 GeV	
±W boson	160 GeV	
±T quark	340 GeV	

[1] https://home.cern/science/physics/matter-antimatter-asymmetry-problem

separation. The time frame of the strong force is about 10^{-24} seconds, about 10^{20} Planck ticks, so the original single Planck pixel was now $2^{10^{22}}$ pixels! My MagicNumber app, at its limit, tells me that $2^{10^5} = 10^{30,103}$ (pixels, $10^{28,00}$ lightyears) so the size of the universe today, which has only doubled in size a few times since the end of inflation, must be truly stupendous.

The energy between the separated quarks was enormous, and this crashed into spacetime as the hot Big-Bang, slowing the rate of inflation of spacetime to the sedate rate we see today. It was this energy that crashed into a plasma of every kind of particle that cooled in the first three minutes to photons and a one-in-ten-billion smattering of hydrogen and helium.[1]

To wax poetical for a moment of summary: Our Universe started with a seed which, guided by the Logos, developed naturally until it flowered into the just-right Universe we exist in today.

The only survivors from this freezing out as the universe cooled was a huge number of photons—added to each time a class of matter-antimatter annihilated—and a minute excess of matter—still unexplained—on the order of 1 matter particle to 100,000,000,000 photons of gamma-ray light. The rounding of numbers in the Biblical "Let there be Light" is now quite understandable.

Most of the matter (~75%) was in the form of hydrogen-1 nuclei (protons) with the surviving neutrons all caught up in helium-4 (~25%) along with deuterium (H-2) in the very small fraction (0.0001%) seen today, all of which was created in the Big-Bang. For while stars 'excrete' all other elements, they 'eat' deuterium avidly. The ratio of relic deuterium (H-2) to hydrogen (H-1) found today is a sensitive measure of the nuclear processes and timeline of the Big Bang 13 billion years ago—small differences in the timeline of the Big-Bang result in great changes to this ratio.

Note that the fusion of H-2 into He-4 is so energy producing that a liter of seawater with its 0.00001% H-2 has more free energy than that in a liter of gasoline. While this fact might make Arab sheikhs swoon in panic, we are as yet incapable of accessing this gift from God.

After the first three minutes of dynamic changes, the Universe settles down to stately expansion and cooling, and the hot plasma condensed into atoms, then massive stars as gravity pulled things together. These condensations of matter all had one measure in common, that of *escape velocity*.

---

[1] Guth, Alan H. (1997). *The Inflationary Universe: The Quest for a New Theory of Cosmic Origins.* Basic Books. pp. 233

Escape Velocity

Here on Earth, if we throw a ball into the air, it will ascend a certain height, stop and then fall back to Earth. (This being a thought experiment that ignores air resistance.) The harder you throw, the higher it will ascend before gravity slows it to a halt and then the ball starts its fall back.

If, however you were Superman and could throw a ball at Mach 33—thirty-three times the velocity of sound—it would escape the Earth and go into orbit, contradicting the maxim that, "What goes up must come down." Note: this does not apply to rockets that use thrust from their exhaust on the way up so can start off majestically slowly. The escape velocity from the Earth's surface is 11.2 km/sec (7 miles/second, 25,000 mph) which is why it is so difficult to get up there into orbit without continuously expending energy.

The Sun is 333,000 times more massive than the Earth, but its radius is much, much greater so the surface is much further from the center. These two parameters cancel somewhat so the escape velocity from the Sun's surface is only 55 times greater than Earth's at 617.5 km/sec. The equation is simple, the escape velocity is proportional to the ratio of mass over radius (distance from center). The velocity increases as the mass increases or the radius decreases

$$V = \sqrt{\frac{M}{R}}$$

In its old age, however, when all its fuel is exhausted, our Sun will shrink into a White Dwarf held up solely by the quantum Pauli Exclusion Principle that forbids electrons to be forced into sharing the same state. The mass will be roughly the same but the radius will be about Earth's size so the cancellation no longer applies and the escape velocity from this mummified Sun will be 6,450 km/s (14,000,000 mph).

If the gravity of a more massive sun is too intense, however, the electrons give up the fight and merge with the protons to create neutrons. Such a neutron star the mass of the Sun would be about the size of Brooklyn with a radius of just ~9 miles. The star would still be held up by the Pauli Exclusion Principle, but now by the quantum impossibility of neutrons being forced to share the same state. The gravity at the surface is now so intense that the escape velocity would be ~150,000 km/s, that is half the speed of light.

Gravity is inexorable, however, and given enough mass the radius can shrink to where the escape velocity reaches the speed of light. Even light cannot escape. This is called the Schwarzschild radius, and every mass has an associated radius with a lightspeed escape velocity. The Sun has a Schwarzschild radius of approximately 3 km (1.9 mi), whereas Earth's is only about 9 mm (0.35 in) and the Moon's is about 0.1 mm (0.0039 in).

Any material body that collapses within its Schwarzschild radius becomes a Black Hole because it is invisible to all forms of electromagnetic radiation—radio waves, visible light, UV, X-rays, gamma rays—it absorbs them all with no possibility of escape.[1]

Sizes of Black-Holes

Astronomical observation has uncovered two major sizes of Black Holes in the Universe: the stellar and the supermassive (with recent intimations of intermediate sizes).

Stellar-sized Black Holes are the result of the collapse of a massive star. Even a neutron has its limits, and if a neutron star is too massive its neutrons dissolve into a plasma of quarks and it shrinks within its Schwarzschild radius and disappears from view. If there is anything close by, however, like a cloud of gas or a companion star, material can fall into the gravitational well, be heated to high temperature, and radiate X-rays profusely from well outside the Schwarzschild radius. These have masses in the range 5-50 times the mass of the Sun.

In our galaxy, such an X-ray source in the constellation Cygnus is the first such source widely accepted to be a black hole. It was discovered in 1964 during a rocket flight and is one of the strongest X-ray sources seen from Earth even though it is more than 6,000 lightyears distant. It is only noticeable because it is sucking mass from its massive companion, a blue giant star. This stolen matter is heated as it falls and radiates X-rays before disappearing into the gullet of the Black-Hole.

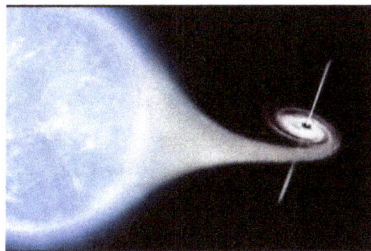

Supermassive-sized Black Holes were discovered when attempts to understand the quasars, the centers of galaxies that were so luminous that they were observed at distances of 10 billion light years. They were pouring out energy at a rate that mass must be being converted to pure energy at a rate of a few suns-worth every day. Furthermore, this mass-to-energy was happening within a diameter of a just tens of light minutes, smaller than the orbit of Mars about the Sun.

The only plausible mechanism for such a prodigious release of energy was the aggregation of matter into a massive Black Hole, which theory suggested could convert almost 50% of the in-falling matter into pure energy

---

[1] Caveat: Given trillions of years, a Black Hole could evaporate by Hawking radiation that, according to theory, should be emitted by a Black Hole's surface.

that could radiate away while the rest plunged through the event horizon and out of view.

Such monsters—active or inactive—are found at the centers of all (?) galaxies, and seem to be an inherent step in the primordial condensation of galaxies out of the tenuous hydrogen-helium gas of the early universe. They act as an axis around which a galaxy rotates. Our galaxy has a central Black Hole that, from our point of view, is near the border of the constellations Sagittarius and Scorpius. It is called Sagittarius A*, and is a bright and very compact astronomical radio source at the center of the Milky Way, 35,000 light years from us. Unfortunately this gargantuan pivot around which our sun rotates every 200 million years or so, is obscured from visibility by great clouds of interstellar gas. We can only agree with Dylan Thomas' Mrs Dai Bread One, "Ach, the mean old clouds!"[1]

The mass of this central Black Hole is four million times that of the Sun. Its Schwarzschild radius is 27 million miles so it would fit comfortably within the orbit of Mercury at 35 million miles. Our 'home' Black Hole, however, is not a champion, size wise. The current champ is in a galaxy 3.5 billion light-years away, containing an ultra-massive black hole with a mass estimated at 18 billion Suns!

Structure of Black-Holes

First we consider the structure of stellar Black-Holes. Current thinking has that once neutrons collapse beyond the Schwarzschild radius there is nothing to resist the inexorable pull of gravity to the very center. The density rises without bound, the distortion of spacetime becomes extreme resulting in a *singularity:*

> A one-dimensional point which contains a huge mass in an infinitely small space, where density and gravity become infinite and space-time curves infinitely, and where the laws of physics as we know them cease to operate.[2]

For reasons I cannot fathom, such theories neglect a basic principle of physics: the closer particles are squeezed together, the hotter they get. The shrinking of the distance   apart is proportional to the third power of the shrinking of the radius—halve the radius, the contents are squeezed eight-times closer.

---

[1] Dylan Thomas, "Under Milk Wood"

[2] https://www.physicsoftheuniverse.com/topics_blackholes_singularities.html

Temperature is a statistical measure of the average kinetic energy of particles; some have less energy than the average, some have more. The kinetic energy is easily converted into electromagnetic radiation, the thermal radiation that is characteristic for any temperature. The higher the temperature, the more intense is the radiation, the energy density increasing as the fourth-power of the temperature. Doubling the temperature raises the density sixteen-fold; a ten-fold rise in temperature increases the density ten thousand times; a million-fold rise in temperature increases the energy density a trillion-trillion-fold.

As noted, the second well-established fact is that photons with sufficient energy can transform into particle pairs of matter and antimatter. The photon has only to have at least the rest mass-energy of the particle pair.

For instance, an electron-positron pair has a mass-energy of ~1MeV. So any volume in the collapsing Black-Hole where the photons have greater energy than this, will also be filled with an equal number of electrons and positrons. The photons turn into matter-antimatter as fast as the matter-antimatter annihilates into photons, and they are all in thermal equilibrium.

The mass-energy of nucleon-antinucleon pairs is ~2GeV and above this energy level protons, antiprotons, neutrons and antineutrons are created in abundance and enter into the thermal equilibrium.

The quarks in a neutron are ~10^{-15} meters apart. When the radius of a neutron star shrinking within its Schwarzschild radius is reduced to 1/10th, this separation is reduced to 10^{-45} and the temperature rises precipitously.

At a temperature of 10 billion K, a strange phenomenon is known to happen: the energy of the photon is sufficient to create electron/anti-electron pairs, which can recombine as photons or as neutrino/antineutrino pairs. The energy density of all this added to the imploding quarks is equivalent to 10^{31} nucleons/liter.

As the temperature continues to rise towards the center, at 10^{14}K protons/antiprotons and neutrons/antineutrons appear from the photons, up again and a plethora of all the particles in the quantum zoo appear, including the top quark and the Z boson, and the energy density greatly increases. At 10^{29}K the energy density is equivalent to packing all the contents of a billion visible universes (10^{88} nucleons) into a tiny volume.

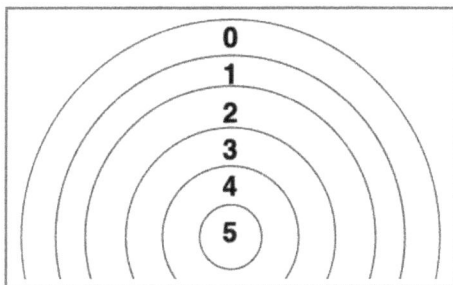

At 1 TeV the weak bosons are traveling at essentially lightspeed and the weak and electromagnetic interactions merge into a single force, a temperature called the GUT freezing point.

As the temperature soars to the maximum possible Planck temperature of 10^{33} degrees, all the forces and particles merge into one, a temperature that is called the TOE freezing point. The energy density is now a trillion trillion trillion trillion trillion trillion trillion trillion times that of neutron stars: immense, but not infinite.

It is obvious that a layer that is 1,000 times denser than a second layer will have no problem excluding it from entering. With this in mind, we see that the black hole has an onion-like structure where each inner layer is thousands of time denser than an outer layer and so excludes it.

entity	photon E	Big-Bang	Black-Hole
± electrons	1 MeV		
± muons	211 MeV		
± protons	1.88 GeV		
± neutrons	1.89 Gev		
±C quark	2.5 GeV		
± tauon	3.5 GeV	freeze out	created in
± B quark	8.5 GeV		
Z boson	95 GeV		
±W boson	160 GeV		
±T quark	340 GeV		

As might be imagined, the outward pressure is such as to halt the inward crush of gravity. The resultant structure is onion-like with regular nucleon density in Zone 0 just inside the Schwarzschild radius at temperatures in the hundreds of millions degrees. Zone 1 is a higher temperature and density of electrons/anti-electrons supporting Zone 0. Zone 2 is at a higher temperature and density with nucleons/anti-nucleons etc supporting Zone 1. Zone 3 at a higher temperature and density of all the quantum zoo of particles and antiparticles supporting Zone 2.

This is probably sufficient for stellar mass Black Holes. Note that this sequence is basically stages of the Big Bang in reverse, heating up rather than cooling down.

In the supermassive variety things probably progress to Zone 4 and beyond where predicted particles such as the ultra massive X and anti-X join in all supporting Zone 3. As temperatures rise to the Plank Temperature (10^{32}K), the separations are reduced to the Plank length (10^{-35}m) and the density rises to the Plank density (10^{93} g/cm3) the conditions approach that of the speck of False Vacuum with which the creation of the universe

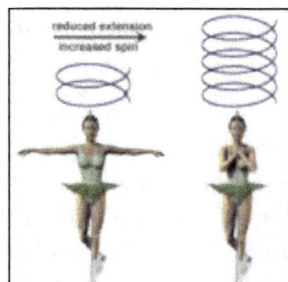
reduced extension
increased spin

began.[1]

This is Zone 5 and it is more than capable of supporting a billion Suns of mass in Zone 4 and be the center of the onion. Gravity has now crushed spacetime, reversing the inflation of the Big-Bang, it has 'deflated' spacetime into a speck of False Vacuum, creating, so to speak poetically, into the seed of another Universe.

In this view of a Black Hole there is no infinite singularity, so it does not suffer the problem that arises once one infinite concept is accepted, others also appear. An example is the conservation of angular momentum which depends on the extension from the axis of the spinning object. A decrease in extension of a spinning body causes the rate of spin to increase.

This is experienced by a spinning skater; as the arms are drawn inwards, the skater's spin rapidly increases to keep the angular momentum constant.

All of the neutron stars that we know of are spinning, some at a rate of a thousand times a second. As a spinning neutron star collapses towards a singularity, in the current view, its rate of rotation has to increase. As the star shrinks to zero extension and infinite density its rate of rotation increases to infinity. This concept is as troubling as the concept of an infinite density.

The layered view presented here does not suffer from these infinities.

Any kind of experimental detail that could distinguish between the two views, however, will have to wait until technology enables the coalescing of two Black Holes to be observed, when the event horizon is ripped and we might get a peek at what the inside looks like. As the super massive black holes are probably formed out of the coalescence of smaller ones, the finely detailed study of quasars will probably be able to settle the question about the inner structure, if any, of Black Holes.

Purpose of Black-Holes

For science in general, when a theory gives an infinite answer, it is usually considered incorrect. Current theories of Black-Holes involve infinities. While God and the abstract realm in which He exists is certainly infinite, humans are not equipped to handle the infinite. A singularity, with its infinite density and distortion of spacetime coupled with "where the laws of physics as we know them cease to operate," is not amenable to human reason. While extremely speculative, nothing in our suggestion is considered infinite or beyond understanding.

---

[1] Guth, Alan H. (1997). The Inflationary Universe: The Quest for a New Theory of Cosmic Origins. Basic Books. pp. 233

In terms of Purpose, Black-Holes do play an important role in Creation as axes for galactic rotation. The Unification Thought perspective is that all things were created by God, via the Logos, with a role in fulfilling the purpose of creation, a wonderful home for His children. There is no suggestion that the role of the physical realm is temporary as a place for humans to create themselves and multiply before shedding their flesh and entering the spiritual realm (as discussed in another chapter).

This is where current physics disagrees with UT. We have already mention the demise of the Sun. Our Sun is ~5 billion years old and it is middle aged. Estimates are that H-burning will continue for another 5 billion before He burning takes over and the Sun becomes hostile to life on Earth. This universe we inhabit has a 'use by' date attached. The interstellar hydrogen used to create new stars is gradually being depleted so the formation of new stars is slowly declining. At some point, there will be no new stars, and all the stars will slowly run out of fuel.

In *Unification Thought*, however, God intended the human race to be ever expanding, so logically at some time humans are going to follow their Heavenly Parent's example and create a fresh new Universe. The main ingredient—the Logos—is already here and is not going away. The only other necessity is a speck of False Vacuum, and if this perspective is correct, there are plenty of them around, safely stored away at the center of supermassive Black Holes.

All some highly advanced technology of our distant descendants will have to do is liberate a speck from its cocoon. Then under the guidance of the Logos, it will inflate into a brand new universe—not inside this one but into the infinity of the abstract realm—ready to be a new home for the human race. While this sounds like science fiction, who knows what can be accomplished after a billion years of technological advance—it is, after all, only 200 years between Volta making a frog's leg twitch and the iPhone and the internet!

I'd like to claim to be a first in this field of Deep Future, but a similar, if somewhat erroneous, scenario is to be found in Isaac Asimov's SciFi tale, "The Last Question." The death of the Universe is followed by a Humanity-Computer amalgam finding the Answer to the Last Question—spoiler alert—and proclaiming, "Let There Be Light!" Teenage thrills in the 1960s!

This is a suitably positive thought to end this discussion.

HYOJEONG ACADEMIC FOUNDATION OF THE ARTS & SCIENCES

T he Hyojeong Academic Foundation of the Arts & Sciences (HJA) has the ambition to establish a universal science in Unification Thought. When accomplished, this will greatly contribute to Rev. Sun Myung Moon's ideal of uniting all sciences along with promoting a science centered on absolute values in this value-neutral era of science.

HJA promotes the unity of inductive and deductive methods, and the development of universal science within Unification Thought. I explore this philosophical endeavor in my treatise, *A New Renaissance: Systematizing the Academic Studies of Godism*[1] that sheds new light on epistemology, ontology and metaphysics in regard to the foundation of universal science within Unification Thought. The aim of HJA is to establish Unification Thought as a science by securing the validity of Unification Thought theories and its empirical objectivity.

Even though Kant, one of the greats, himself held that his view of the mind and consciousness were inessential to his main purpose, some of the ideas central to his point of view have come to have an enormous influence on his successors. Some of his ideas are now central to cognitive science, for example. Professionals debate Kant's model as a whole and the claims in it that have been influential. His claims about consciousness of self specifically are important to Unification Thought. Indeed, even though he achieved remarkable insights into consciousness of self, many of these insights next appeared only about 200 years later, in the 1960s and 1970s.

The dominant model of the mind in contemporary cognitive science is Kantian, but some of his most distinctive contributions have not been taken into account.

According to some philosophers, object recognition proceeds in three stages: first feature detection, then location of features on a map of locations, and then integration and identification of objects under concepts. This compares directly to Kant's three-stage model of apprehension of features, association of features (reproduction), and recognition of integrated groups of under concepts.

This is mirrored in the three stages necessary to 1. go from an insight at the formation stage to 2. The development and maturation of this ending in 3. The perfection and completion as an addition to human thought.

---

[1] http://iaap.info/index.php?option=com_content&view=article&id=105&Itemid=1131

This work is at the formation stage of God & Science merging into a Unified Science. I hope it will inspire others to move into the second and third stages.

This effort by HJA is not unique; there have been many efforts in history to establish a universal science. This history, perhaps, got started with Plato's theory of Ideas as the basis of all studies. The Platonic doctrine of recollection or anamnesis, is the idea that we are born possessing all knowledge and our realization of that knowledge is contingent on our discovery of it. Whether the doctrine should be taken literally or not is a subject of debate.

Plato's thinking was based on the concept that the soul is trapped in the body, while Unification Thought has the soul being created by the combination of God's Love and the actions of physical body. The soul once lived in "Reality", but got trapped in the body. It once knew everything, but forgot it. The goal of Recollection is to get back to true Knowledge. To do this, one must overcome the body, a position expressed many times by Reverend Moon, the instigator of Unification Thought. In Unification Thought this pre-existence resonates with the concept of the Original Image—in the Inner hyungsang of God's structure—embracing all things, including the image of each one us in a not-fallen state that we can grow into as we mature and fulfill the human portion of responsibility.

This anamnesis doctrine implies that nothing is ever learned, it is simply recalled or remembered. In short it says that all that we know already comes pre-loaded on birth and our senses enable us to identify and recognize the stratified information in our mind. Many scientists have expressed their puzzlement as to just where a crucial insight came from as it popped into their mind and their ideas fell into order. It is my hope that some of the ideas expressed in this book will resonate with those in your mind and simulate a host of other related ideas.

The effort to find unity continued Aristotle's effort to establish the structural system of a science based on formal logic. After the medieval stagnation, Descartes attempt to systematize geometry as a universal science, Husserl's efforts to materialize the ideology of a universal science from the logical structure of pure consciousness, etc., are groundwork based on subjectivity. On the other hand, a view that finds its roots in Bacon's inductive method, Galileo's and the Enlightenment's view of mathematization of nature as an ideology based on universal science, and attempts by the Unity of Science movement in Vienna Circle based on establishing physics as a standard of universal science, etc., are construed as foundation work based on objectivity.

Complementary dualities that work together constructively are a fundamental principle in Unification Thought. An illustrative example is to be found in the history of the scientific understanding of electricity and magnetism. Michael Faraday (1791-1867) had a very intuitive approach and invented the concept of electric and magnetic fields and developed the theories of electromagnetic induction, amongst other contributions. His complement was James Clerk Maxwell (1832-1879) who applied advanced mathematical operations to Faraday's insights and deduced a set of equations that fully encapsulated electromagnetism—the Maxwell Equations that are as valid in quantum physics as they were in classical physics—and discovered the electromagnetic nature of light.

This book will also deal with the confusion that often occurs when different languages impede communication. Religion has its lexicon—here it is that of the *Divine Principle*, elucidated by Rev. Moon that, as promised in the Bible, speaks plainly of the Father. Philosophy has a different lexicon—here it is Unification Thought, systematizing DP with philosophical considerations as initiated by Sang Hun Lee. Science has yet another lexicon developed over the ages, with fluid concepts that can change as experimental methods become more sophisticated and results insist that ideas change. The Quantum revolution with its disruption of a scores of classical concepts is the science lexicon used in this work.

Translating between the three areas is necessary as in many cases they are using different words for the same thing. For example, the terms: Principle of Creation, the Logos, Natural Law while seemingly different are actually dealing with the same concept: the mathematical patterns that govern Nature. This book has a chapter about that.

Another example are the dualities: mind and body, *eidos* and *hyle,* wave and particle which, while appearing diverse, are actually dealing with the same aspect of reality. There is chapter on that.

An apparently insurmountable chasm is one of the disciplines makes claims that are anathema to another. An example is the religious concept of a spiritual realm that coexist side-by-side with the physical realm. For many scientists this is just wishful thinking, a solace to mortality, akin to a belief in fairies and elves. Yet, as another chapter will explore, while a spirit realm was inconceivable in classical science, modern science is much more amenable to the concept.

Another example is in the area of evolution, where science claims that sophisticated development occurs solely by chance an accident. Religion, on the other hand, asserts that evolution occurs by input from the Logos. There is a chapter on that dichotomy.

The concept of free will is central to religion with its focus on sin and redemption. This is a difficult concept in science where determinism is deeply rooted. The first chapter takes on this disjunction of ideas.

Yet another apparent stumbling block to a unified picture of the world is the origin of mankind. Science describes this as apes gradually developing into hominids, and hominids gradually developing into humans. Religion has it that the change was abrupt, a saltation event with the birth of Adam and Eve, the first human couple. The chapter on this topic discusses how science is gradually moving towards the view of religion.

This book takes a fresh look at relations between the two great magisteria of Religion and Science, that are "non-overlapping " in the thought of Stephen Jay Gould. While many of the founders of science were theists, the general consensus these days seems to be that an understanding of either God or Science can have nothing in common, in fact, are in opposition and contradict each other.

This perspective is challenged in this HJA book which briefly examine areas of thought where the two perspectives are reconciling. Naturally, this is not intended to be the last word on the matter but as a stimulant to further efforts to promote unity of perspectives on this most fundamental existential question of God or no-God.

This work is a step forward in establishing a logical structure for universal science that embraces religion, philosophy and science in a coherent picture of the world we live in.

Sung-Bae Jin
Hyojeong Academic Foundation of the Arts & Sciences

INDEX

protostomes 4, 14
Ptolemy 57
punctuated equilibrium 82
Pythagorean 16, 62
pythons 66

Q

quark 16, 21, 23, 47
quark-antiquark 24

R

radians 30
Radio 21, 28, 73, 86, 88
random-chance-and-accident 83
randomly 3, 10, 39, 79
Read-Only 76, 77
receptor 83, 93
recombination 81, 82, 87
relativity 28, 60
religion 1, 3, 8, 10, 27, 54, 64
Renaissance 42
renaturing 38
replicate 13. 91
reproduction 74, 77
reptiles 14, 65
resonance 20, 40, 49
reverse-transcriptase-like 77
ribose 88
ribosome 37, 73, 74, 75
ribozymes 76
RNA-viral 77
RNAs 75, 89, 91, 94
Rubisco 37

S

S-quark 24
satellite 35
Schrodinger 33, 98
Seattle 70, 97
secreting 93
seed 16, 72
Selectric 89
serendipity 45
sewer 35
sexual 81

shrink-and-turn 31
silicon 50, 87
silver 92
Sirius 35
slit experiment 7, 35
snail 93
snake 65
sodium 33
solar-system 50
somatic 81, 83
space-time 62
Speciation 10, 80, 84
spectrum 40, 58, 86
sperm 83
sphincters 94
spin 19
spinach 75
spirit-body 64
spiritual 2, 5, 63, 67
spliceosomes 90
splicing 76, 90
stability 33, 49, 72, 82
Standard Model 62
stars 45, 50, 57, 85
stellar 48
sub-subsystems 5
suborning 94
substrate 39, 41, 93
subsystem 9, 71
sugar 93, 96
sulfur 50
Sun 36, 24, 50, 57
super-symmetrical 62
Supercluster 58
supercoiled 80
supernova 50, 59
supersymmetric 61, 64, 65
SUSY 61
symbiotic 55
symbols 87
Symmetry 19, 28, 61
synapse 93, 95

T

T-quark 24
tachyon 60
tardyon 61
tau-neutrino 24
tauon 24

CPSIA information can be obtained
at www.ICGtesting.com
Printed in the USA
LVHW081506041222
734563LV00021B/266